SPACE SHUTTLE LOG
The First 25 Flights

Gene Gurney and Jeff Forte

AERO
A division of TAB BOOKS Inc.
Blue Ridge Summit, PA

The authors are grateful to Althea Washington of the National Aeronautics and Space Administration for her assistance in supplying the photographs for this book, and to John Lawrence, Program Analyst in the Office of Space Flight, NASA Headquarters, for his technical expertise.

REF
TL
795.5
.G87
1988

FIRST EDITION
FIRST PRINTING

Copyright © 1988 by TAB BOOKS Inc.
Printed in the United States of America

Reproduction or publication of the content in any manner, without express permission of the publisher, is prohibited. No liability is assumed with respect to the use of the information herein.

Library of Congress Cataloging in Publication Data

Gurney, Gene.
  Space shuttle log.
  Includes index.
  1. Space shuttles—History. I. Forte, Jeff.
II. Title.
TL795.5.F67 1987   629.45'009   87-30713
  ISBN 0-8306-8390-9 (pbk.)

Questions regarding the content of this book
should be addressed to:

Reader Inquiry Branch
TAB BOOKS Inc.
Blue Ridge Summit, PA 17294-0214

Front cover photographs courtesy of NASA.

Contents

	Foreword	v
	Preface	vii
	Prologue	ix

FLIGHT
1	COLUMBIA: Flight of Firsts	*1*
2	Columbia's Second Mission	*15*
3	A Busy and Successful Test Mission	*27*
4	Simulating Operational Flight	*39*
5	First Operational Flight	*49*
6	CHALLENGER: A Second Orbiter	*61*
7	Four Guys and Sally Ride into Space	*71*
8	Testing a New Communications System	*87*
9	Spacelab	*99*
10	Untethered Spacewalks	*107*
11	Solar Max Repair	*117*
12	DISCOVERY: A Third Orbiter	*129*
13	Largest Crew Yet	*143*

14	Satellite Pickup and Delivery	*153*
15	First Secret Military Mission	*161*
16	A Flyswatter and a Senator	*167*
17	7 Humans, 2 Monkeys, 24 Rats	*177*
18	An International Crew	*189*
19	False Start, Fine Ending	*203*
20	Fixing Leasat 3	*217*
21	ATLANTIS: Fourth Orbiter, Secret Mission	*231*
22	West Germany's Spacelab	*237*
23	Building Space Structures	*245*
24	Last Complete Flight	*259*
25	The Challenger Disaster	*267*
	Epilogue	*284*
	Index	*289*

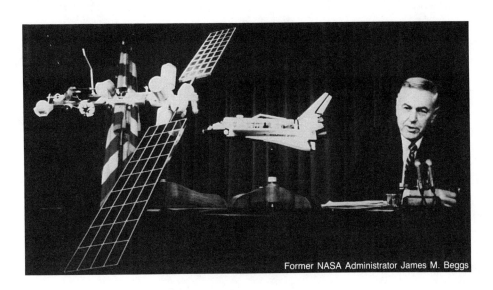
Former NASA Administrator James M. Beggs

Foreword

FROM THE FIRST AMERICAN ORBITAL FLIGHT OF JOHN GLENN IN FRIENDSHIP 7 ON FEBRUARY 20, 1962 to the first flight of Columbia on April 12, 1981 spans a period of enormous progress in America's space program. The United States landed 12 human beings on the moon, the first humans to walk and work on an extraterrestrial body. American automated spacecraft explored the far reaches of our solar system; we visited Mercury and Venus, put a lander on Mars, and sent our Pioneer and Voyager explorers beyond the asteroid belt to the primordial planets of Jupiter, Saturn, Uranus, and Neptune. It was indeed an exciting two decades.

But with the first flight of the space shuttle, America began a new era of space exploration and exploitation. For the first time we had the means to work in space much as we do here on Earth. Our astronauts could live in a controlled environment with sufficient elbow room to conduct experiments and observations of the Earth and the universe unparalleled since the beginning of the space age. Moreover, we were learning how to live and work on extended-duration flights which were the progenitor of a more permanent occupancy of space.

It was an exciting period, for almost five years we built on our experience and flew more and more. We logged more time and flew more than twice as many individuals than in all the previous history of human spaceflight in the United States. American women took a prominent role in the shuttle program, with the flight by Sally Ride on Flight Seven and other women crewmembers on many subsequent flights. Several foreign astronauts flew with the shuttles, and we initiated a program of inviting non-astronaut crewmembers from the general public. Two chairmen of NASA's committees in the Congress, Senator Jake Garn and Congressman Bill Nelson, flew with the shuttle.

In December of 1985 the program began to falter, and on January 28, 1986, Challenger tragically was destroyed with its entire crew. Thus ended the first set of shuttle flights. I say "first" because I fervently believe that there will be many more flights to come.

When NASA restarts the shuttle flights, it will be with a significantly different machine and a new set of objectives. But the experience we have gained will enable us to move quickly to regain our momentum in space science and exploration as we move toward the space station era in the 1990s.

Gene Gurney is uniquely positioned to chronicle the years of the shuttle flights. He has served his country and NASA for over 40 years and has written extensively on the space program. I have enjoyed reliving the exhilaration and the disappointments of these memorable flights.

There is a lot of NASA's history in *Pericles*, called *Lost and Found* by students of Shakespeare. If I may quote Helicanus, the Lord of Tyre:

> Sure, all's effectless; yet nothing we'll omit
> That bears recovery's name. But since your kindness
> We have stretched thus far, let us beseech you
> That for our gold we may provision have,
> Wherein we are not destitute for want,
> But Weary for the staleness.

James M. Beggs
Administrator, NASA
1981 - 1986

Preface

THE SPACE SHUTTLE WAS THE ANSWER TO THE QUESTION, "WHAT DO WE DO NEXT IN manned space?" At the end of the Mercury program, the question was asked if manned spaceflight should continue, and for some period of time the answer was no. The same question was again asked for the end of the Apollo program, but in the early 1970s, when it was decided that the United States would remain in the manned space business, the serious development of the space shuttle started.

The space shuttle was developed to be an economical means of getting in space to do things that were not affordable otherwise. It was to open the door to doing new things in space. The space shuttle is often compared to other launch vehicles, which is incorrect because it is more than a means to place payloads in orbit. Much comparison has been done between it and expendable launch vehicles regarding dollars per pound of payload in orbit and such, but the shuttle, being a manned vehicle, can do much more. It has the utility of being a space laboratory and repair shop, and is able to return, as well as deliver, payloads.

The vehicle is a very complex machine that requires many systems to work every time. The addition of man to the system was to make it more reliable as well as add flexibility to its mission. Making the first flight manned, something never done before in spaceflight, was strongly questioned as being too dangerous. But studies indicated much additional complexity was required to do some of the task if the man was removed. If the computers that fly the vehicle failed to operate properly, there were situations when the crew could take over and correct the situation. Certainly the development of an automatic landing capability would have been expensive and time-consuming—to perform a task that the flight crews have shown they can perform very well.

The first flight of the shuttle proved a number of concepts that were under research and development for years. The concept of a winged recoverable and reusable spaceflight vehicle had been under study since the early 1950s. A new reusable surface insulator (tile) thermal protection system was used successfully for the first time. A new dual-cycle, high-specific-impulse, main liquid propulsion system, which is bottleable and reusable, was flown for the first time. A flight control system using fly-by-wire digital computers was used to add stability in the many different flight regimes up to Mach 26. This represented a state-of-the-art system. Many other new technology systems were used, and the first flight represented a proof of the design concept.

The shuttle demonstrated its utility much earlier in its flight program than expected. The failure of the Solar Max satellite provided an excellent opportunity to demonstrate the capability of repairing and reservicing a satellite on orbit. This, of course, saved considerable money in eliminating the requirement to build and launch another satellite. When solid motor upper stages failed on communication satellites being launched by the shuttle, it was possible on a later mission (after delivering other payloads to orbit) to pick up and return the expensive satellites and have them repaired for later launching. When another satellite experienced a failure in the firing circuit that ignites its solid motor, a later shuttle flight was able to correct the problems, and the satellite was successfully placed in the higher orbit it required. These relatively early unique missions showed what could be accomplished only by a *manned* shuttle.

Dr. Milton A. Silveira
Chief Engineer, NASA
1983 - 1986

Prologue

AS THE TIME GAP WIDENS BETWEEN THE LAST SHUTTLE FLIGHT AND THE NEXT ONE, THE set of 25 flights that occurred between 1981 and 1986 begins to take on the characteristics of a separate and distinct era in spaceflight history.

When the shuttle does fly again, enough changes in the shuttle's propulsion, safety, and even management systems will have taken place that the United States will have, in effect, a new Space Transportation System.

To chronicle the events of this most recent era in U.S. manned spaceflight history is the purpose of this book. It ends with the tragic Challenger accident but is not about the Challenger accident. Others have probed and will continue to probe what went wrong on January 28, 1986. Rather, here, each of the 25 flights is detailed from launch preparation, launch, flight operations, and right on through to the landing.

From the reader, no special background or knowledge about spaceflight is assumed. Great care has been taken to eliminate much of the technical jargon that easily pops up in a field of human endeavor so dependent on technology. What is offered is an honest attempt at an objective, historical record of a period that can be called the "First Shuttle Era." There will be other eras to come because there was so much left undone during these few, short years. The world only got a glimpse of what could be possible in attempting routine access to space.

The Presidential Commission on the Space Shuttle Challenger Accident, headed by former Secretary of State William P. Rogers, spent four months poring over every detail of the Challenger accident and, in effect, the entire shuttle program. In its June 6, 1986 report to the President, the Commission concluded:

> The Commission urges that NASA continue to receive the support of the Administration and the nation. The agency constitutes a national resource that plays a critical role in space exploration and development. It also provides a symbol of national pride and technological leadership. The Commission applauds NASA's spectacular achievements of the past and anticipates impressive achievements to come. The findings and recommendations presented in this report are intended to contribute to the future NASA successes that the nation both expects and requires as the 21st century approaches.

Mission Number:	STS-1 **Orbiter:** Columbia
Crew:	John W. Young, Commander (*left*) Robert L. Crippen, Pilot (*right*)
Launch Prep:	Orbiter Processing Facility: 610 days Vehicle Assembly Building: 35 days Launch Pad: 105 days
Launch from KSC:	April 12, 1981; 7:00 A.M. (EST) Attempt on April 10 scrubbed because of timing skew in orbiter general-purpose computer system.
Mission Duration:	2 days, 6 hours, 20 minutes, 52 seconds **Orbits:** 36
Distance Traveled:	933,757 miles
Landing:	April 14, 1981; 10:21 A.M. (PST) Edwards Air Force Base, California
Wheels-Down to Stop:	8,993 feet
Returned to KSC:	April 28, 1981
Mission:	Major space shuttle systems were tested successfully. Orbiter sustained some tile damage from overpressure wave created by the solid rocket boosters. 16 tiles lost and 148 damaged.

FLIGHT 1: COLUMBIA

COLUMBIA:
Flight of Firsts

"TODAY OUR FRIENDS AND ADVERSARIES ARE REMINDED THAT WE ARE A FREE PEOPLE capable of great deeds. We are a free people in search of progress of mankind."

With those words President Ronald Reagan welcomed home the two-man crew of Columbia on April 14, 1981. On that day, John W. Young and Robert L. Crippen made a perfect landing on the hard-packed bed of Rogers Lake in California's Mojave Desert, after a nearly flawless voyage in space. This flight ushered in a new era of manned space travel for the United States after an absence of half a dozen years.

This was a flight of firsts: the first launching of an airplane-like spacecraft with wings and landing gear; the first time solid rockets were used to launch a manned spacecraft; the first recovery of boosters for reuse; and the first time that any American spacecraft was put into orbit without prior unmanned testing.

According to NASA plans, STS-1 was to be the first of four "test" flights. It would carry no payload except for a data-collection and recording package. The basic flight objective was to demonstrate safe launch into orbit and return to landing of the orbiter and crew. A second objective was to verify the combined performance of the entire shuttle vehicle—orbiter, solid rocket boosters, external tank—up through separation and retrieval of the spent rocket boosters.

Launch Preparations

Assembly of the space shuttle's "stack" for the STS-1 mission began in December 1979. The stack consists of the external tank and the twin solid rocket boosters, to be erected on the orbiter on a mobile launcher platform in the Kennedy Space Center Vehicle Assembly Building's High Bay 3.

The space shuttle orbiter Columbia arrived at Kennedy Space Center from Dryden Flight Research Center in California aboard the 747 Shuttle Carrier Aircraft on March 24, 1979, and was immediately moved into the Orbiter Processing Facility for systems checkout and the completion of installing its thermal protection system, which is the tile covering on the entry areas of the orbiter.

Arriving by barge from the Michoud Assembly Facility in New Orleans, the huge external tank underwent processing in the Vehicle Assembly Building during the summer of 1980. In the fall of that year, the three components of the vehicle—orbiter, boosters and external tank—were put together, and various checkouts of the mechanical and electrical connections between them began. The assembled space shuttle, aboard its mobile launcher platform, was moved the 3½ miles from the Vehicle Assembly Building to Pad A on December 29, 1980 to undergo final processing for launch. At the end of February, a successful 20-second Flight Readiness Firing of Columbia's three main engines was completed. After the firing, steps were taken to repair a small portion of the external tank's insulation, which had come loose. In March, astronauts John Young and Bob Crippen went through a "dress rehearsal" countdown and simulated liftoff. Other major testing continued and cleared the way for the countdown and launch scheduled for April 10, 1981. The countdown on April 10 proceeded normally until 20 minutes prior to launch when the orbiter's general purpose computers malfunctioned. Analysis and testing indicated that the primary set of general purpose computers provided incorrect timing to the backup flight system and caused the delay.

At 7:00 A.M. on the morning of April 12, 1981, Columbia's three main liquid-hydrogen-fueled rocket engines and two solid rocket boosters generated nearly 6½ million pounds of thrust to lift the approximately 4½ million-pound space shuttle from Launch Pad 39A at Kennedy Space Center, Florida. Rising on a pillar of orange flame and white steam, the shuttle cleared its over 300-foot-high launch tower in just six seconds and reached Earth orbit in about 12 minutes. The solid rocket boosters and external fuel tank were jettisoned just prior to orbit.

"Man, that was one fantastic ride," exclaimed Bob Crippen, who was on his first spaceflight, as his heartbeat rose from 60 to 130 per minute.

John Young, a veteran of four previous spaceflights including an Apollo moon landing, had a heartbeat rise of from only 60 to 85. Later the 50-year-old Young said he was excited too, and jocularly added, "But I just can't make it go any faster."

Orbital Operations

Young and Crippen changed their orbit from its lower original elliptical orbit to a higher circular one by firing their Orbital Maneuvering System. Later they raised their orbit to 175 statute miles above the Earth. They tried out all systems and conducted a number of engineering tests. They checked the computers, the jet thrusters used in orienting Columbia, and the opening and closing of the large cargo bay doors. Aside from allowing the release of satellites into space, opening the cargo bay doors was critical to releasing the heat that built up in the crew compartment. Closing them was necessary for return to Earth.

Young and Crippen documented their flight in still and motion pictures. One view of the cargo bay, which was telecast to Earth, revealed that all or part of 16 heat-shielding tiles was lost, probably due to stresses of launch. The missing or damaged tiles were from the two pods on the tail section that house Columbia's rockets. The loss was not considered critical as these pods would not be subjected to intense heat. However, there were areas on the spacecraft's underside, nose, wings, and tail where the frictional heat would be generated by entry into the atmosphere that would reach 3,000 °F. As it turned out, the more than 30,000 remaining tiles all stuck firmly through reentry. The tiles were made of a material that shed heat so readily that they could be red hot on one side and cool enough to touch on the other. In addition, the tiles would not burn away, as other heat-shielding materials did on previous spacecraft.

Return to Earth

About an hour and a half before they were to land on April 14, the two astronauts fired their rockets for a short burst to reduce speed from their orbital velocity of around 17,500 miles per hour. When they fired their rockets, they were over the Indian Ocean, beginning their descent into the landing field at Edwards Air Force Base in California. They fired their attitude control thrusters to turn Columbia right-side-up relative to Earth and nose forward to be in the proper profile for landing. They fired thrusters again to pitch Columbia's nose up at an angle so that the brunt of atmospheric reentry pressures and temperatures would be taken by Columbia's broad, well-protected underside.

After completing the fiery reentry into the atmosphere, Columbia's computers changed its steering from rockets to a more airplane-like configuration that used the rudder and elevons (a combination of ailerons and elevators commonly used on delta-winged craft) to guide Columbia through the atmosphere. Columbia continued its drop toward Earth like a very fast glider. Air drag caused the craft to steadily lose speed as its altitude dropped.

According to plan, the two pilots guided Columbia to its landing strip on the bed of Rogers Dry Lake in the Mojave Desert of California. They banked sharply left and looped back, lining up for a final approach. Then they touched down at a speed of about 215 miles per hour, which is nearly twice that at which commercial jetliners ordinarily land. The touchdown marked the successful conclusion of STS-1, 2 days, 6 hours, 20 minutes, and 52 seconds after liftoff from Florida.

"What a way to come to California," exulted Crippen.

As soon as Columbia stopped, it was surrounded by a convoy of vehicles carrying specialists who took measures to remove dangerous concentrations of explosive and poisonous gases in Columbia's cargo bay and in space around its engines. They ventilated the entire craft and withdrew residual fuel from the engines. After an hour the crew was then permitted to leave Columbia and go to a waiting medical van. John Young came out first. Before going to the van, he carefully inspected Columbia's exterior and kicked a tire in the traditional test pilot's sign of acceptance of a vehicle that carried him through a successful flight. His inspection completed, he smiled broadly and gave a thumbs-up victory gesture.

Spaceship Columbia, the Space Transportation System first orbiter, with its prime crew, John Young (left seat) and Robert Crippen (right seat), going over the checklist during a power-up mission simulation in the Orbiter Processing Facility.

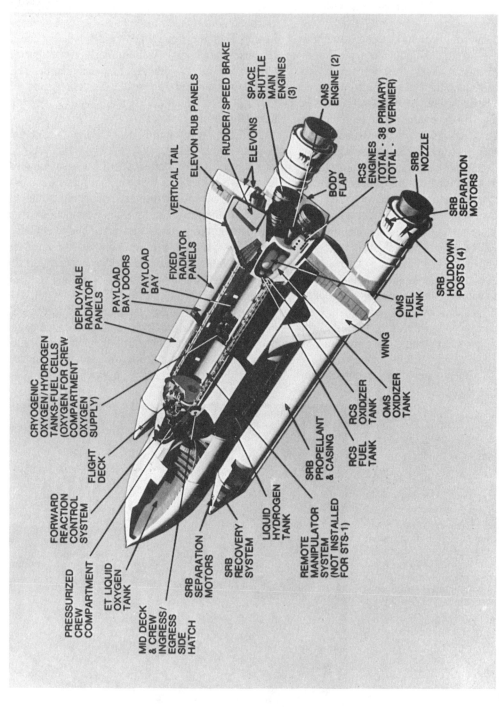

This "cutaway" artist's concept exercises some artistic license to reveal systems of the major components of a space shuttle vehicle. With its payload-bay doors open here, the shuttle's cryogenic supply station and the deployable radiator panels can be seen. In reality, the cargo bay panels would never be open while the orbiter is attached to the solid rocket boosters (SRBs) and external fuel tank (ET). The thick-bodied, delta-winged aerospace craft is 37 meters long, has a span of 24 meters (120 feet by 80 feet), and weighs about 75,000 kilograms empty (165,000 lbs.). Its payload bay, 18.3 meters long and 4.6 meters in diameter, can carry maximum payloads up to 370 kilometers altitude or smaller loads up to 1,110 kilometers (230 miles to 690 miles). It can bring payloads of 14,575 kilograms (32,000 lbs.) back to earth, and it can carry out a variety of missions lasting 7-30 days.

Contingency landing operations are rehearsed using a mockup of the orbiter's cockpit. The mockup is moored in the Turning Basin at Complex 39 to simulate a shallow-water ditching of the orbiter. Helicopters deploy Air Force and Navy divers who are specially trained to rescue the astronauts from the cockpit.

Astronauts Crippen (left) and Young (right) pose in front of the shuttle orbiter Columbia as it sits atop NASA's 747 jet transporter. The astronauts were on hand to welcome the orbiter to the Kennedy Space Center. Columbia was ferried 2,400 miles from the Dryden Flight Research Center, California. After experiencing technical delays and bad weather, the aircraft touched down at the Shuttle Landing Facility at 11:03 A.M. (EDT) on March 24, 1979.

COLUMBIA: **Flight of Firsts**

Vice President George Bush is advised by astronauts Young (center) and Crippen (right) as he tries out the contoured launch seat and non-activated controls of the orbiter Columbia on Launch Pad 39A. Bush is garbed in the required white coverall issued in the "white room" entrance to the orbiter's cockpit.

This dramatic night shot of the space shuttle and its reflection in the waters of the extensive lagoonal system adjacent to Launch Pad 39A was taken on the evening prior to the Flight Readiness Firing of Columbia's main engines. The 20-second firing on February 20 was a milestone procedure in flight preparation of the shuttle.

Wings of fire and feathers provide dramatic contrast as Columbia and one of the graceful inhabitants of the ecologically rich KSC area both lift free of Earth's restraints. A remote camera at Launch Pad 39A captured this view as the maiden flight of STS-1 began.

COLUMBIA: Flight of Firsts

One of the solid rocket boosters used in the launch is "dewatered" at sea on April 13. The expended booster casings, lowered to the Atlantic Ocean by parachutes, fill partially with water and float in a vertical mode until special nozzle plugs are inserted in the rocket nozzle throats, and the water is pumped out. Here the casing is rising from the sea, just before toppling over into a horizontal or "load" float mode for towing. The casings are towed to Cape Canaveral Air Force Station for reprocessing and eventual reloading by the Morton Thiokol Corporation.

A frustrum from one of the two solid rocket boosters is recovered at sea on April 13 by one of the two recovery ships, UTC Freedom and UTC Liberty, specially built for the purpose. The frustum, located just aft of the nose cone, contains the main parachute that lowers the expended rocket casing into the sea for recovery and reuse.

Astronaut Young mans the commander's station in Columbia during the 36-orbit STS-1 flight. A loose-leaf notebook with flight activities data floats in the weightless environment. Young is wearing a three-piece constant-wear flight suit.

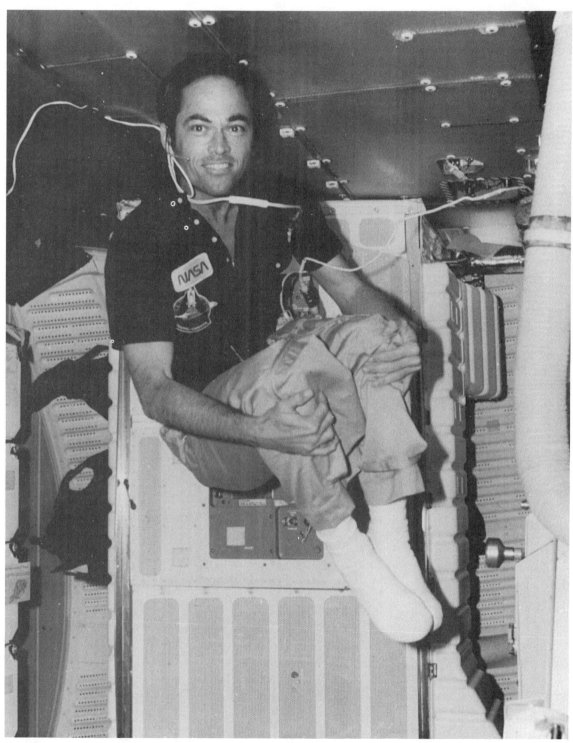

Astronaut Robert L. Crippen, pilot for STS-1, takes advantage of zero gravity to do some unusual acrobatics aboard Columbia.

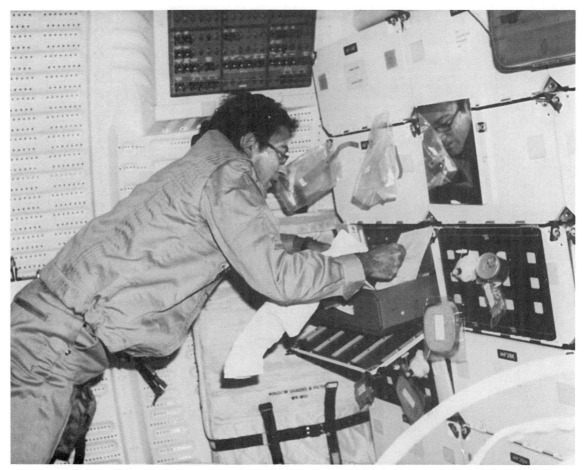

Astronaut Young, cleans off his razor after shaving. A food tray can be seen mounted on a locker door at right.

This frame shows the cargo bay of Columbia. The thermal tiles that were learned to be missing during the flight's first day came from the left and right Orbital Maneuvering System (OMS) pods. Especially clear is the area of missing pieces on the left pod. The developmental flight instrumentation package (DFI) can be seen in the lower portion of the frame.

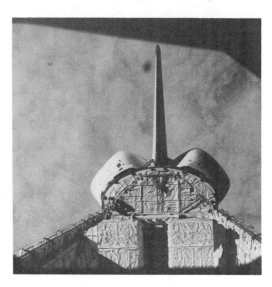

Completing the first full test of the Space Transportation System, the orbiter Columbia is seen here on its final approach to Rogers Dry Lake Runway 23 at Edwards Air Force Base in California, April 14, 1981.

President Ronald Reagan poses with astronauts Young and Crippen, NASA Acting Administrator Dr. Alan Lovelace, and Vice President George Bush in the Oval Office of the White House during a ceremony in which the astronauts received NASA's Distinguished Service Medal. Young also received the Congressional Space Medal of Honor, and Lovelace was presented with the President's Citizens Medal.

Mission Number:	STS-2 **Orbiter:** Columbia
Crew:	Joseph H. Engle, Commander (*left*) Richard H. Truly, Pilot (*right*)
Launch Prep:	Orbiter Processing Facility: 103 days Vehicle Assembly Building: 21 days Launch Pad: 74 days
Launch from KSC:	November 12, 1981; 10:10 A.M. (EST) First set for October 9 but delayed by spill of nitrogen tetroxide during loading of forward reaction-control system. An attempt November 4 was scrubbed when countdown computer called for a hold in the count because of apparent low reading on fuel-cell oxygen-tank pressures. During hold, high oil pressures were discovered in two of the three auxiliary power units that operate the hydraulic system. Filter replacement required delaying launch until 7:30 A.M. on November 12. There was a further delay of three hours to replace multiplexer/demultiplexer. Modifications of launch platform to overcome overpressure problem were found to be effective.
Mission Duration:	2 days, 6 hours, 13 minutes, 12 seconds **Orbits:** 36
Distance Traveled:	933,757 miles
Landing:	November 14, 1981; 1:23 P.M. (PST) Edwards Air Force Base, California
Wheels-Down to Stop:	7,000 feet
Returned to KSC:	November 25, 1981
Mission:	The flight was cut from its planned duration of five days because of failure of one of three fuel cells that produce electricity and drinking water. The Remote Manipulator System was tested for the first time. Mission scientists were satisfied with the data from "earth-looking" experiments in payload bay. No tiles were lost, but about a dozen were damaged.

Columbia's Second Mission

WITH THE SECOND LAUNCH OF THE SPACE SHUTTLE IN NOVEMBER 1981, A NEW ERA IN SPACE travel began. For the first time a spacecraft was flown more than once.

The orbiter Columbia lifted off from Florida for what turned out to be a shortened mission. Originally scheduled to last for more than five days, the mission was cut short by trouble in one of Columbia's fuel cells. These cells converted hydrogen and oxygen into electrical power for the spacecraft and into drinking water for the crew. With one of three fuel cells malfunctioning, mission safety rules called for STS-2 to be reduced to a minimal mission lasting 54 hours, circling the globe 36 times. After some deliberation, test managers decided to cut the mission short, in this case by more than half.

During post-flight ceremonies at the Dryden Flight Research Center, Commander Joseph Engle said that he was "kind of heartbroken to come back early," but added, "We had a good flight, a fun flight, and hope we have contributed something meaningful to spaceflight."

Launch Preparation

In September, launch of the shuttle was significantly set back when nitrogen tetroxide oxidizer being pumped into Columbia's attitude-control rocket tanks spilled onto the heat-shielding tiles. The oxidizer acted like a solvent, dissolving the cement beneath the tiles. A total of about 380 tiles had to be carefully cemented back onto the shuttle.

Just before the launch on November 4, the countdown computer called for a hold in the countdown. During the hold, high oil pressures were discovered in two of Columbia's three auxiliary power units that operate the craft's hydraulic system. The hydraulic system swiveled Columbia's rocket engines during launch and operated its aerodynamic controls and landing gear after atmosphere reentry when the spacecraft becomes an aircraft. The oil filters in two of the auxiliary power units were found to be clogged and were replaced. This set the launch back to November 12 at 7:30 A.M.

However, unexplained problems arose with Columbia's units that display data on the craft's condition to both the crew and ground control. New units were removed from the orbiter Challenger, which was currently under construction in California, and flown to Kennedy Space Center. This activity delayed the launch two and one-half hours.

Liftoff of STS-2

On the first shuttle liftoff, shock waves from the rocket blast had bent several of Columbia's fuel tank supports. On the second launch, this problem was solved by a newly developed water deluge system installed at the bottom of the launch pad. Some 300,000 gallons of water were sprayed into the rocket exhaust. This dampened the shock wave by more than 75 percent and prevented damage to Columbia on its second trip into space.

The booster rockets and the external propellant tank detached and fell into the ocean as planned. It took two days to recover the spent boosters, however, because of high winds and waves in the area.

Despite the smoothness of the flight, the failed fuel cell proved too worrisome to warrant attempting the planned mission of 83 orbits. Among the considerations were that, if another fuel cell failed, Columbia would not have enough electric power to both bring itself in and acquire vital data such as heating of the craft during atmosphere reentry. Such data were lost on the first shuttle flight. The judgment of flight test managers was that the risks of a longer flight far outweighed possible gains.

The only other instances of fuel cell problems in the U.S. space program occurred on Gemini 5 and Apollo 13, and these were due to the fuel tank or fuel line—not the cell itself.

Orbital Operations

Although their time in space was substantially reduced, Commander Engle and Pilot Richard Truly crammed extra work into the available time, providing abundant data for orbital flight test engineers and other experimenters. They achieved more than 90 percent of the objectives set for the STS-2 flight.

Among their major goals was the first test in space of a Canadian-built Remote Manipulator System. The system was comprised of a huge mechanical arm, operating from Columbia's payload bay and guided by controls on Columbia's control deck. The system was designed to deploy payloads into orbit and retrieve them, as well as for other freight-handling activities in space. The arm could even be used to reach around and inspect various external parts of Columbia. It had its own lighting system and closed-circuit television so that the crews operating it had a close-up view of the tasks they were performing.

The mechanical arm was jointed like a human arm is, joined at the shoulder, elbow, and wrist. Fully extended, it was 50 feet long and just over a foot across. Despite its immense size, it was made of sturdy, lightweight materials giving it a weight on Earth of only 900 pounds. The comparison with the human hand ended with the wrist. The "hand" part was called an "end effector" and consisted of a snare wire device that could be tightened around grapples attached to the payloads.

Engle and Truly operated the arm in all of its modes, ranging from fully automatic, in which it was programmed in advance by computer to perform a series of operations, to fully manual, in which it was operated directly from a control panel that bypassed the computer.

Engle and Truly did not move any payloads with the arm because this test did not call for it. They did let the arm's camera picture them holding a sign saying "Hi, Mom!" A vital part of their test was swinging the arm back onto its cradle pedestals along the left side of Columbia's payload bay and latching it there.

Because the first shuttle flight was a "test" flight, Columbia carried a complement of instruments to record the performance characteristics of its various systems. Valuable engineering data was accumulated by a flight information recorder. In STS-1, the recorder malfunctioned, losing the data. This loss was serious because the orbital flight tests were designed to find and remedy defects in the shuttle system and qualify it for routine operation.

In addition to gathering engineering data, Columbia in STS-2 conducted experiments that would contribute to such fields as prospecting for oil, gas, coal, and minerals, locating promising ocean fishing grounds, understanding how gravity affects plant growth, forecasting thunderstorms and other severe weather, and measuring air pollution. Goals of the experiments included: determining whether Columbia could provide a stable platform for conducting Earth surveys, testing advanced techniques and instruments to survey Earth from space, and gathering data about Earth's resources and environment. STS-2 met these objectives despite the shortened mission, although it did not provide all of the data about Earth's environment and resources that were originally called for.

Most of the instruments for these experiments were carried in an engineering model of the Spacelab pallet installed in Columbia's payload bay. The Spacelab pallet not only provided a structure on which to mount experiments, but also such support as electrical power. The pallet was designed for experiments requiring a broad field of view or direct exposure to the space environment.

The Earth survey experiments needed an unobstructed view of Earth, making the Spacelab pallet a welcome advantage. Data from these experiments were accumulated on Columbia rather than radioed to Earth, so that Columbia's data transmission capabilities could be devoted to flight test information. The experiment data were provided to the experimenters after Columbia landed.

As in any flight test program, the orbital flight tests were designed to identify and resolve unanticipated problems. Some of these problems resulted in launch delays. The test program, among other things, was designed to shorten turnaround time.

A break in the planned work of the flight occurred on the day after launch, when President Reagan visited the Mission Control Center at Houston. He chatted with the two astronauts, Engle and Truly, and jokingly asked them about hitching a ride with them to California. On a serious note, he told them that the whole nation and the world were watching them with great pride.

Return to Earth

Engle and Truly followed the same course in returning from orbit as did Crippen and Young on the first flight: retrofire over the Indian Ocean, enter the atmosphere over the western Pacific Ocean, and transition to aerodynamic controls in the atmosphere to land as an aircraft.

This time Columbia was guided to its landing site by a microwave-scanning-beam landing system on the site. This time also, Engle, a former Air Force test pilot, banked the craft and put it through other maneuvers to test how it handled under stress. Most of the descent, however, was in automatic mode; at level off, Engle took the controls. Plans for a crosswind landing were dropped because of 25-knot winds, too high for safety. Engle made a perfect touchdown despite extensive cloud cover that made the Edwards Air Force Base site marginal for landing.

As in STS-1, shortly after Columbia landed, measures were taken to remove explosive or poisonous gases from its payload bay and along its surface. They also ventilated the craft. After the immediate vicinity around Columbia was determined to be safe, Engle and Truly left the spacecraft and stepped down the ladder at just after 5:00 P.M. (EST) on November 14.

A gradually brightening morning sky provides a backdrop as STS-2 is pictured on its rollout and journey from the Vehicle Assembly Building to Launch Pad 39A. The powerful Crawler Transporter carried the 1,255,000-kilogram (2,766,000-pound) shuttle and the 3,735,000-kg (8,234,000-pound). Mobile Launcher Platform on their 3.5 miles journey at an average speed of less than one mile per hour.

This photograph of STS-2 crewmembers Engle (left) and Truly was made when they were serving as the backup crew for STS-1. They were in the Johnson Space Center's motion-base shuttle mission simulator in the mission simulation and training facility. Many hours of training were logged in this simulator in preparation for the autumn mission of the two astronauts.

The shuttle orbiter Columbia, undergoing launch processing for its second voyage into space, is gently lowered down toward the solid rocket boosters and external tank for mating. The operation began a little more than three months after Columbia returned to KSC following its highly successful maiden flight.

Columbia's Second Mission

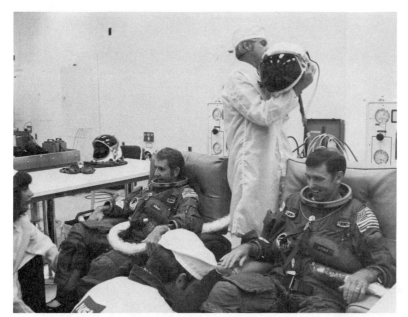

Suit-up operations took place in KSC's Operations and Checkout Building before the crew boarded a van at 4:48 A.M. on November 4 for their ride to the launch pad.

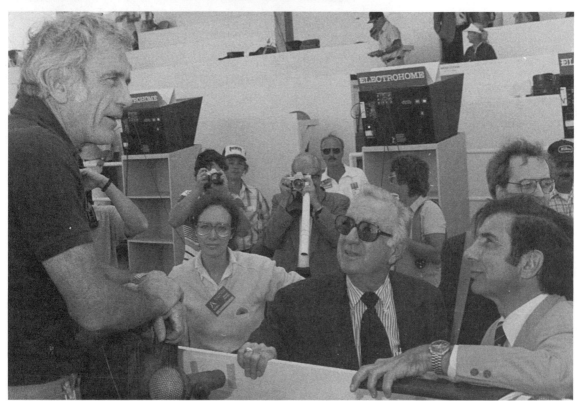

Donald K. "Deke" Slayton (left), orbital flight test manager for the space shuttle program, chats with two prominent members of the press corps, Walter Cronkite of CBS (center) and the late Jules Bergman of ABC, following an STS-2 launch-status press briefing on November 3. This launch was "scrubbed" the following day at T minus 31 seconds.

Columbia climbs toward space for a return visit after its 10:10 A.M. liftoff from Complex 39's Pad A. On its second mission, Columbia carried a payload of science and applications experiments and a Remote Manipulator Arm—a mechanical arm designed to handle cargo in orbit.

Columbia's Second Mission 21

Astronauts on STS-2 heard a woman's voice during their flight—a new occurrence for NASA space missions. Mission Specialist/Astronaut Sally K. Ride, working closely with the Remote Manipulator System (RMS), joined the ranks of NASA's capsule communicators (CAPCOMs) during the operation of the RMS in Earth orbit.

President Ronald Reagan gets a laugh from NASA officials in Mission Control at Johnson Space Center, Houston, Texas when he jokingly asks astronauts Engle and Truly if they could stop by Washington en route to their California landing in order that he might come along. The STS-2 prime crewmen were in their next-to-last day in Earth orbit when the space-to-ground conversation took place. NASA Administrator James M. Beggs is standing, second from right.

Dust flies as Columbia touches down Saturday afternoon November 14, 1981 at Edwards Air Force Base in California.

Columbia's Second Mission 23

Astronauts Engle (left) and Truly (center) complete an inspection of the space shuttle Columbia with George Abbey, Director of Flight Operations for Johnson Space Center. The inspection immediately after landing Saturday afternoon, November 14, 1981, showed that none of the thermal heat tiles was lost, alleviating the concern following the first mission in April 1981.

After completing its second successful mission, Columbia begins its return flight to Florida aboard the Boeing 747 Shuttle Carrier Aircraft.

Columbia's Second Mission

Mission Number:	STS-3 **Orbiter:** Columbia
Crew:	Jack R. Lousma, Commander (*left*) Charles G. Fullerton, Pilot (*right*)
Launch Prep:	Orbiter Processing Facility: 70 days Vehicle Assembly Building: 14 days Launch Pad: 34 days
Launch from KSC:	March 22, 1982; 11:00 A.M. (EST) Launch was delayed one hour by ground-support equipment problem.
Mission Duration:	8 days, 5 minutes **Orbits:** 129
Distance Traveled:	3.3 million miles
Landing:	March 30, 1982; 9:05 A.M. (MST) Northrop Strip, White Sands, New Mexico Landing site changed from Edwards Air Force Base, California, to Northrop because of wet conditions at the Edwards facility and delayed one day because of high winds at Northrop.
Wheels-Down to Stop:	13,810 feet
Returned to KSC:	April 6, 1982
Mission:	Continued testing of space shuttle systems for qualification for operational flight. Extensive testing of the Remote Manipulator System. Measurement of thermal response of orbiter in various attitudes to the Sun. Nine space science experiments flown, plus monodisperse latex reactor, electrophoresis test, heflex bioengineering test, and induced environment contamination monitor. First student experiment flown. Problems included space sickness, malfunctioning toilet, thermostat difficulty, and unexplained static interfering with crew sleep in flight on ascent. One auxiliary power unit registered overheating but functioned properly on descent. Three communications links were lost March 26. Thirty-six tiles were lost and 19 were damaged.

A Busy and Successful Test Mission

EACH SHUTTLE MISSION HAS ELEMENTS THAT MAKE IT UNIQUE FROM PREVIOUS FLIGHTS, even though for the third flight it was a subtle difference. Columbia's third mission into space began about four months after completion of its second mission—substantially less than the seven months between the first and second flights. The STS-3 launch was on the day originally scheduled, March 22, 1982—a first for the shuttle. It was delayed just one hour from the planned 10:00 A.M. launch.

Orbital Operations

STS-3 was the third of four planned orbital test flights designed to eliminate as many difficulties as possible before the shuttle was considered "operational." In STS-3 some problems proved bothersome, but most turned out to be minor. About seven minutes after launch, a sensor flashed a message that one of the three Auxiliary Power Units (APUs) on Columbia was overheating. The APUs provide hydraulic pressure to swivel the rocket engines during the launch phase and operate the rudder and elevons during atmosphere reentry. The orbiter could have operated adequately using only two of its three APUs during ascent and reentry modes (they were not used in orbit). During return to Earth on this mission, however, all three units operated properly.

On the day after launch, astronauts Lousma and Fullerton discovered that Columbia had lost a few dozen of its 31,000 heat-protection tiles. The loss of these particular tiles would not endanger the spacecraft during reentry. An inspection of Columbia after it landed revealed the loss of 36 full tiles and parts of 19 others.

Early in the mission the crew encountered space sickness, a balky toilet, and temperature control and radio static problems that interfered with sleep. A thermostat difficulty kept the cabin either too warm or too cold. Whenever Columbia passed over a certain area of Asia, the crew's radio headsets crackled with static, disturbing the crew who were trying to sleep. The static was attributed to a powerful radar station. Most troubles were corrected by the third day, and the astronauts went about their tasks in good health and high spirits for the remainder of the eight-day mission.

On the fourth day of the mission, three of the communications links between Columbia and Earth were lost due to equipment malfunctions. Besides the remaining high-power communications link, a backup FM radio and UHF voice circuit were still available. Loss of the radio links reduced data transmission from Columbia to the ground stations but did not threaten the safety of the mission.

Space Science Project

The mission's space science segment was considered a pioneering effort because many of the experiments were designed to gather information about Columbia's interactions with the space environment and to help set standards for future experiments. One experiment was designed to find out how the presence of Columbia in space affected certain ionized gases and magnetic-field lines. Other experiments were designed to indicate how the delicate measurement of celestial and terrestrial objects would be affected by the dumping of water and release of certain gases from the shuttle.

Another experiment Columbia carried was designed to provide information about buildup of condensation on instruments, caused by volatile materials expended by Columbia. Tests of the dynamic, acoustic, and thermal properties of the shuttle's environment were made in this experiment.

On this mission Columbia carried experiments to monitor X-rays emitted by solar flares; to gain more information about formation of lignin (the stiffening agent in plant stems) in the nearly gravity-free environment of space; to study the relationship of plant height in space to initial soil moisture content; to gather information about the frequency, mass, and chemistry of micrometeorites; and to find how flying insects behave in a nearly gravity-free environment.

Among the insects that Lousma and Fullerton photographed with a motion picture camera were honeybees and velvet-bean-caterpillar moths. This experiment was submitted by Todd E. Nelson, a Minnesota high school student. His experiment was one of 10 national winners from 1,500 proposed experiments submitted in the 1980-81 NASA National Science Teachers Association Shuttle Student Involvement Project competition. The NASA/NSTA project was designed to stimulate the study of science and technology in the nation's secondary schools.

To prepare for future low-cost science experiments in space, a payload canister for small, self-contained experiments called Getaway Specials was developed. Getaway Specials were experiments developed by industry, schools, individuals, and other organizations and were scheduled for flight on a space-available basis.

Medical and Materials-Processing Experiments

One experiment flown on the third mission employed a process to separate kidney and blood cells. Known as "electrophoresis," the process uses a small electric charge to separate a solution containing different types of cells. On Earth, the electrophoresis process produces undesirable convection currents in fluids, currents that tend to prevent separation of cells that closely resemble one another. The value of this experiment, which was flown repeatedly during the 25 flights, was its potential for commercial processing of chemicals and pharmaceuticals in space.

The materials-processing experiment of STS-3 involved the development of uniform-sized latex spheres. Experimenters wanted to determine whether weightlessness could help produce uniform spheres, which is difficult in Earth gravity. These spheres have various medical applications, including measurement of pores in the intestines, eye research, and cancer treatment.

Thermal Testing

A major STS-3 goal was thermal testing of Columbia to determine its performance in the temperature extremes of space. First, the tail was pointed toward the Sun for 28 hours. Later, the nose was pointed toward the Sun for 80 hours, and finally, the top of the ship—with the cargo bay doors open—was directed at the Sun for 28 hours. Between these exposures, Lousma and Fullerton rolled the craft for periods of 3 to 10 hours' duration to equalize external temperatures. While subjecting Columbia to thermal stresses, they opened and closed the payload bay doors. Following prolonged exposure of the open bay doors to intense cold away from the Sun, the doors would not latch properly after closing. Latching was completed normally after exposure to the Sun heated the top of the cargo bay.

The crew started and restarted Columbia's orbital maneuvering engines and operated the huge mechanical arm of the Remote Manipulator System. They developed considerable expertise in using the manipulator arm to grasp an experiment called the Plasma Diagnostics Package. They moved this unit around outside and inside the payload bay and returned it to its stowed position. These maneuvers were accomplished despite a short circuit that blacked out the manipulator arm's television wrist camera. They guided the 50-foot arm using only a secondary one—the elbow camera. This marked the first time the arm was used to take cargo out of the spacecraft. In later flights, the arm was to prove invaluable in satellite repair work.

Lousma and Fullerton also fired their attitude-control rockets to roll and pitch Columbia while they used the manipulator arm. These movements caused no noticeable arm tremors or loss of grip. As they berthed the arm for the last time, Fullerton called it a "fantastic piece of machinery."

An Unusual Landing

Rogers Dry Lake Bed at Edwards Air Force Base was the primary landing field for the shuttle orbital flight tests. Both previous flights had landed there. However, heavy rains had drenched the lake bed and it was no longer "dry." No one knew how long it would be before the surface would be dry enough to land the shuttle. Northrop Air Strip at White Sands, New Mexico was a hard-packed desert floor and was chosen as the best alternative site for a landing.

Plans called for the STS-3 landing at midday on March 29, 1982. As Lousma and Fullerton were preparing their spacecraft for reentry into the atmosphere, the winds started to blow much stronger at the White Sands landing strip. John W. Young, commander of the first flight, piloted a jet aircraft over the landing area. He measured winds much too stiff for Columbia and observed that a severe sandstorm had cut visibility at the landing side to almost zero. "I think we ought to knock this off," he radioed to Mission Control. "We concur," came the reply.

Just 39 minutes before they were scheduled to fire their braking rockets to descend from orbit, Lousma and Fullerton were "waved off." Lousma landed Columbia the next morning under clear skies and acceptable wind conditions.

Moving the landing site from Edwards to White Sands meant that facilities for processing Columbia after landing had to be set up at White Sands. Equipment and technicians needed for the landing were transported from Edwards to White Sands in 38 railroad cars forming two special trains. Just over 6,000 people viewed the landing at the White Sands site.

James M. Beggs, administrator of NASA, said after the landing, "We've proven once again that, when we want to, we can perform splendid things and make them look easy."

Scheduled for flight aboard STS-3, the Getaway Special Flight Verification Test Canister is shown mounted on the side of the payload bay of orbiter Columbia as the spacecraft underwent preflight preparations in the Orbiter Processing Facility. The test was intended to measure the environment in a two-by-three-foot cylindrical container during an actual mission. The data was analyzed for use by Getaway Special experimenters on future shuttle missions. Officially titled "Small Self-Contained Payloads," the program was offered by NASA to anyone who wished to fly a small scientific experiment aboard the shuttle on a space-available basis. At the time, more than 320 reservations had been made by experimenters from the U.S. and 14 foreign nations.

The two veteran astronaut crewmembers of STS-3 are seen here seated in a shuttle mission simulator at the Johnson Space Center, astronauts Lousma (left) and Fullerton (right). Lousma was pilot for the second of three NASA crews to visit the Earth-orbiting Skylab space station in 1973. Fullerton, though never before having been in space, was a veteran of three free-flights of the shuttle during approach and landing tests with space shuttle Enterprise in 1977.

Lousma (left) and Fullerton (right) look on as student Todd E. Nelson explains his "Insects-in-Flight Motion Study" project. The experiment, sponsored by the Avionics Division of Honeywell, Inc., focused on the flight behavior in zero gravity of two species of flying insects with differing ratios of body-mass to wing-area. Nelson, a senior from Southland Public School, Adams, Minn., was the first of 10 students to have a project selected for flight in the National Space Shuttle Student Involvement Project.

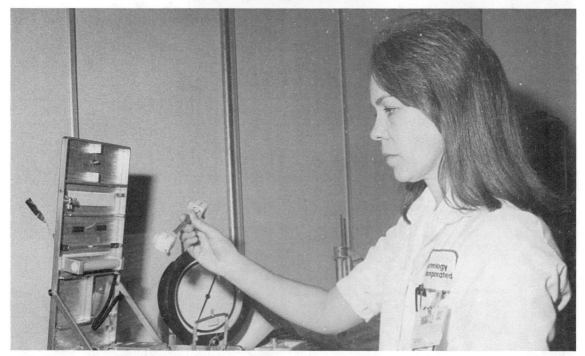

Dr. Marian Lewis, head of the Bioprocessing Laboratory at Johnson Space Center, inspects the Electrophoresis Equipment Verification Test unit to be flown into Earth-orbit aboard STS-3. Designed to separate human kidney cells by using the small negative electric charges on their surfaces while in weightless environment, the process allows culture growth back on Earth of the particular cells containing an enzyme that can dissolve dangerous blood clots. The cultures were expected to reduce costs and increase availability of treatment. The equipment pictured was first flown aboard Apollo/Soyuz and was refurbished for shuttle missions.

A Busy and Successful Test Mission

Illuminated against the early morning darkness by banks of powerful searchlights, STS-3 begins its slow journey, via Crawler Transporter, from the Vehicle Assembly Building to Launch Pad 39A.

STS-3 thunders aloft on twin clouds of fire, beginning a planned week-long mission. Columbia began its third trip into orbit at a fraction of a second before 11:00 A.M. (EST) on March 22, 1982.

A Busy and Successful Test Mission 33

Columbia appears as a tiny flaming triangle with a trail of puffy smoke on this photo taken by astronaut Richard O. Covey from a T-38 chase plane. The T-38 in the photo was taking television imagery.

The UTC Liberty—one of two booster recovery ships in the "NASA Navy"—tows a solid rocket booster through Port Canaveral after recovery operations in the Atlantic east of Cape Canaveral following the launch of STS-3. The Liberty sailed in through the jetties about 6:00 P.M. and was followed by the UTC Freedom about 25 minutes later towing the other solid rocket booster. The Freedom and its sister ship recovered the boosters and related hardware at sea approximately 164 statute miles northeast of Pad 39A.

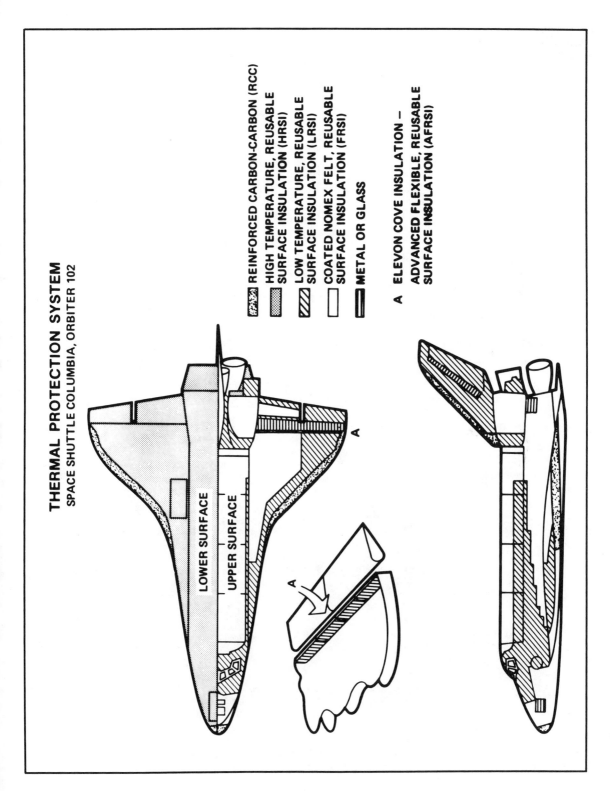

A Busy and Successful Test Mission 35

This picture, taken by astronaut Lousma, shows astronaut Fullerton busy with mealtime preparations aboard the Earth-orbiting Columbia. Positioned in the corner of the mid-deck area, Fullerton holds a beverage in his right hand in an accordion-like squeeze dispenser. Many packages of prepackaged dehydrated foods and other meal items can be seen fastened by velcro to the locker doors and to trays on the locker doors.

Columbia touches down on Northrop Strip at White Sands Missle Range, New Mexico.

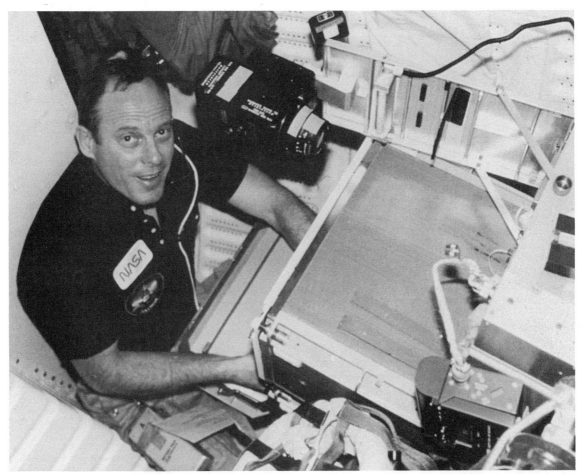

The L-shaped experiment in the right half of this photo was one of a number of scientific experiments which made the trip for NASA's third space mission, along with astronauts Lousma, pictured, and Fullerton. The experiment, making a debut in space (it also flew on the Apollo/Soyuz Test Project in 1975), was designed to evaluate the feasibility of separating cells according to their surface electrical charge. It was a forerunner of planned experiments with other equipment that would purify biological materials in the low-gravity environment of space.

A Busy and Successful Test Mission

Mission Number:	STS-4 **Orbiter:** Columbia
Crew:	Thomas K. Mattingly, Commander (*right*) Henry W. Hartsfield, Pilot (*left*)
Launch Prep:	Orbiter Processing Facility: 42 days Vehicle Assembly Building: 7 days Launch Pad: 33 days
Launch from KSC:	June 27, 1982; 11:00 A.M. (EDT) First space shuttle to be launched on-time with no delays in schedule.
Mission Duration:	7 days, 1 hour, 9 minutes, 39 seconds **Orbits:** 112
Distance Traveled:	2.9 million miles
Landing:	July 4, 1982; 9:09 A.M. (PDT) Edwards Air Force Base, California First landing on a concrete strip (15,000-foot Runway 22)
Wheels-Down to Stop:	8,000 feet
Returned to KSC:	July 15, 1982
Mission:	Final STS research and development flight. Cargo included the first Getaway Special, a Defense Department payload, and the first commercial experiment, the Continuous Flow Electrophoresis System. Mattingly and Hartsfield performed two medical experiments on themselves, operated the manipulator arm to swing the Induced Environmental Contamination Monitor around the orbiter, and took photos of lightning activity in the atmosphere below. The two solid rocket boosters were lost when they hit the ocean, but all other mission objectives were achieved.

Simulating Operational Flight

HAILED AS THE SMOOTHEST AND MOST SUCCESSFUL OF THE ORBITER TEST FLIGHTS, STS-4 was called the "golden spike" by President Reagan, who was referring to the excitement of an earlier episode in American history when the nation linked two sections of transcontinental railroad with a golden spike. "Now we move forward," said the President, "to capitalize on the tremendous potential offered by the ultimate frontier of space."

Arriving back on Earth on the nation's 206th birthday, Columbia ended its fourth mission and final research and development flight amid high hopes for routine access to space.

A flawless June 27, 1982 countdown was achieved despite prelaunch torrential rains and a hailstorm that pitted and drenched Columbia's skin tiles. After reaching orbit, Commander Thomas K. Mattingly and Pilot Henry W. Hartsfield reported that "things look pretty nice . . . we're setting up shop."

The two reusable solid rocket boosters were lost when they plunged into the Atlantic Ocean and sank in about 3,000 feet of water. This was the first time that the boosters had not been recovered. "The only conclusion we can draw," said one of the managers, "is that the main parachutes failed to function."

The thermal testing program, begun on the previous mission, was continued on this flight. Originally planned exposures called for pointing the orbiter's tail to the Sun for 66 hours, bottom to the Sun for 33 hours, and payload bay to the Sun for 5 hours. These periods were altered because hail had damaged numerous tiles, allowing them to absorb water. Engineers were concerned that the water-soaked tiles would freeze and be further damaged. Consequently, they positioned the orbiter in such a way that the sun would dry the tiles by vaporizing the water. Soon after launch, temperature readings from the shuttle's outer surface indicated that the wet tiles had dried completely.

While keeping the underside facing the Sun, the crew opened and closed the graphite-epoxy payload bay doors on the cold side of Columbia. One door failed to close properly during this procedure. When a similar problem occurred during the third flight, it was corrected by rolling the orbiter to heat it evenly, like a rotisserie. By repeating this maneuver, Mattingly and Hartsfield were able to operate the payload bay doors easily. This reassured the crew and engineers on Earth that the warping of the orbiter's aluminum airframe, which was responsible for the problem, was a temporary difficulty and could be solved with a relatively simple procedure.

Zero-G Medical Research

The first experiment by a commercial firm, an electrophoresis test designed by McDonnell Douglas Astronautics Co. in collaboration with Ortho Pharmaceutical, was flown as part of a joint endeavor agreement in which NASA and industry became partners in promoting development of advanced commercial products in space. The companies agreed in advance to make the products derived from such experiments available to the public at reasonable cost after FDA approval. Electrophoresis is a technique used to separate biological materials in a fluid according to their electrical charges as they pass through an electrical field. The process is used to produce many pharmaceuticals. On Earth, gravity makes the solution move around, which limits the purity of the material produced. In space, the almost complete lack of gravity greatly enhances this process. Use of this process was expected to eventually lead to better and lower-cost pharmaceuticals.

In another experiment with many medical and scientific applications, STS-4 carried the Monodisperse Latex Reactor, which performed so successfully on the third flight. Some of the STS-3 experiments were used as "seed" items to test whether larger microspheres could be produced in space. The size to which such spheres can be developed on Earth is limited because of gravity. Production in space would develop potential for microspheres that could be used in calibrating instruments such as electron microscopes and in carrying precise amounts of drugs and isotopes directly to diseased or cancerous tissues.

Mattingly and Hartsfield participated in two medical experiments, both of which were winning entries of the Shuttle Student Involvement Project of NASA and the National Science Teachers Association. One experiment was by Amy Kusske of Long Beach, California and the other by Karla Hauersperger of Charlotte, North Carolina. Mattingly's and Hartsfield's blood and urine were sampled before and after their flight. The astronauts also recorded their food intake and exercise periods. Kusske wanted to determine whether proper distribution of cholesterol in the body in microgravity required strenuous exercise. Hauersperger sought to know whether microgravity reduced chromium levels in the body. A chromium deficiency decreases effectiveness of insulin and can produce diabetes-like symptoms. The biomedical laboratories at NASA's Johnson Space Center analyzed the blood and urine samples for both experimenters.

Getaway Specials

STS-4 also carried the first "Getaway Special," the popular name for the Space Transportation System's Small Self-Contained Payload Program. For as little as $3,000, customers anywhere in the world could purchase space in one of the canisters for scientific and technological experiments. NASA took reservations on a first-come, first-served basis. By the time of STS-4, nearly 350 reservations from individuals, organizations, and governments had already been booked from the United States and 14 other countries. The main requirement for this cargo was that it operate automatically and require little attention in space.

The first Getaway Special was purchased by Gilbert Moore of North Ogden, Utah, for $10,000 and donated to Utah State University. Its nine experiments by university students covered such microgravity study areas as the growth of fruitflies, brine shrimp, duckweed, and algae; the thermal conductivity of an oil-and-water mixture; soldering; alloying; surface tension; and the curing of composite materials.

A defective circuit prevented electrical power from reaching the Getaway Special. After several failed attempts to turn on the power for the experiment, Mattingly and Hartsfield followed a technique devised by NASA engineers described as comparable to "hot wiring" an automobile to start it without an ignition key. Anxious University of Utah students learned on June 29 that their experiments had been turned on and offered accolades for the crew by paraphrasing Apollo 11 astronaut Neil Armstrong, "One small switch for NASA, a giant turn-on for us."

The crew employed the huge manipulator arm in the cargo bay twice to lift and swing the Induced Environment Contamination Monitor (IECM) around the payload area to get information about particles, moisture, and gases in the bay that could affect experiments. The bay was facing the Sun during these experiments to permit maximum release of orbiter contaminants.

Through a special radio hookup, the astronauts spoke to the crowds at the Energy World's Fair in Knoxville, Tennessee, on July 1. "We're talking to you from Columbia," said Mattingly as he addressed fair-goers over the public address system. Mattingly described the view over the Mississippi River Valley and the Gulf Coast. "It's fitting that we land on July 4 and celebrate—ushering in a new era just as our forefathers ushered in an era of democracy over 200 years ago on that same date," said Hartsfield. The broadcast marked the opening of NASA's exhibit at the fair.

On July 3, Columbia came within eight miles of a derelict Soviet upper-stage rocket that had been used to launch a satellite in 1975. The crew never saw the object and they were never in danger of collision. "You'd have to be watching exactly the right place at exactly the right time and not blink," said Harold Draughon, a flight director, who was explaining why the crew did not see anything. The North American Air Defense Command tracks such objects and reported that it was one of more than 4,500 man-made objects circling the globe at the time. The precision track around the globe for shuttle missions precludes any "near misses" with objects in space.

Columbia's reentry into the atmosphere and its return to Earth were planned to be more rigorous than previous flights. This was done to generate more heat to test its thermal-resistant structure and protective tiles. The crew also tested Columbia's ability to automatically stabilize itself after rapidly pitching its nose up and down.

"Hard" Landing

For the first time, the shuttle landed on a surface other than a dry lake bed. The 300-foot-wide and 15,000-foot-long concrete runway at Edwards Air Force Base was used as the landing site for STS-4. On hand for the Sunday morning Independence Day landing were an estimated half-million people, quite a few more than the 6,000 that had witnessed the last landing several months before. At the welcoming ceremonies, Thomas K. Mattingly characterized spaceflight in his own way: "You never get tired of it. It's a new discovery every minute and every day."

After the fourth flight, NASA considered the shuttle to be operational. "We are ready to put the Space Transportation System to work, and it will earn its way," announced NASA administrator James M. Beggs.

The Martin Marietta/Michoud external fuel tank, destined to become part of STS-4, is towed from the Complex-39 barge basin to the Vehicle Assembly Building looming in the background. There it was processed and mated with the orbiter and the solid rocket boosters that make up the total shuttle vehicle.

Inside the Orbiter Processing Facility, a Rockwell International technician prefits an instrumentation tile to Columbia's upper forward fuselage, just in front of the spacecraft's windshield. This particular thermal protection tile is being replaced due to damage, while the other tiles missing here were removed to be densified (strengthened).

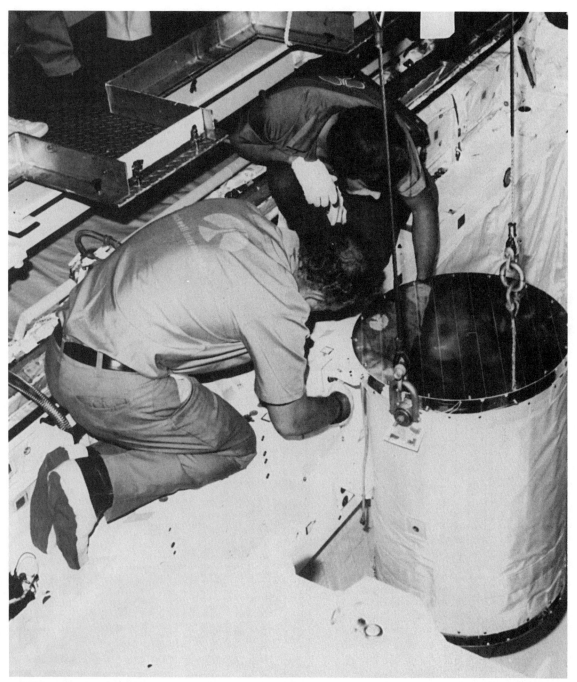

Rockwell technicians attach the first operational Getaway Special to its bridge beam in Columbia's cargo bay. This payload, containing ten scientific experiments prepared by students and graduates of Utah State University and Weber State College in northern Utah, was launched on Columbia's fourth journey into space on June 27, 1982. First in a line of over 300 such payloads reserved by a wide variety of domestic and foreign users, Getaway Special Number One was sponsored by Mr. and Mrs. R. Gilbert Moore of North Ogden, Utah. The original reservation for this payload was made on October 12, 1976.

Simulating Operational Flight

The Crawler Transporter, spanning the two lanes of the turnpike-wide "crawlerway," moves the fourth space shuttle to Launch Pad 39A. Traveling at a top speed of one mile per hour, the crawler made the 3.5-mile journey in just under six hours—a fast trip for a vehicle carrying an 11.7-million-pound load.

Both of Columbia's solid rocket boosters fire to lift the reusable space vehicle off Launch Pad 39A and on its way toward a 241-kilometer (130-nautical-mile) circular orbit. This was the final orbital flight test before STS-5 marked the beginning of the operational era of the Space Transportation System.

44 FLIGHT 4: COLUMBIA

The North Atlantic, southeast of the Bahamas, served as a backdrop for this scene of Columbia's RMS arm and hand-like device (called an end effector) grasping a multi-instrument monitor for detecting contaminants. The experiment was called the Induced Environment Contaminant Monitor (IECM). The small box contained 11 instruments for checking the contaminants in and around the orbiter's cargo bay—substances that might adversely affect the delicate experiments carried on board.

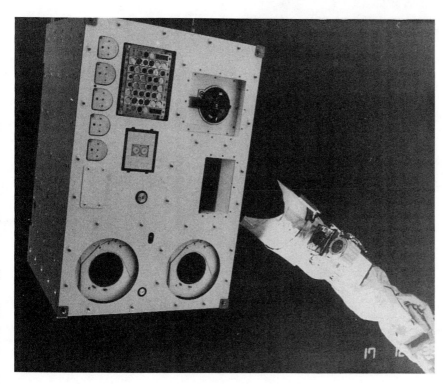

The IECM in a closeup view. The wrist camera on the arm is to the right of the experiment.

Simulating Operational Flight

Exposed through an aft window of Columbia, this scene shows both Florida coasts, with Cape Canaveral prominent at the right center of the frame. Tampa Bay is at lower left.

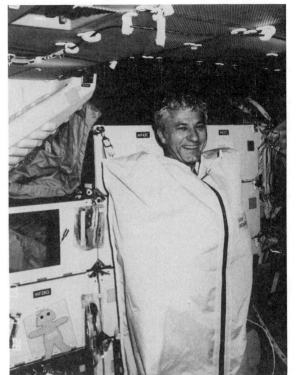

Astronaut Henry W. Hartsfield, Jr., demonstrates the sleeping accommodations on board the Earth-orbiting Columbia. The sleep restraint is located in the mid-deck area of the spacecraft.

STS-4 landed on a beautiful Fourth of July morning at Edwards Air Force Base before rows and rows of visitors from throughout the United States. Commander Thomas K. Mattingly and Pilot Henry W. Hartsfield piloted Columbia to a near-perfect landing, with President and Mrs. Ronald Reagan among the estimated 525,000 guests. Speaking after the landing, the President called the astronauts "pathfinders who lead the way" in "crossing new frontiers every day."

Simulating Operational Flight

Mission Number:	STS-5 **Orbiter:** Columbia
Crew:	Vance Brand, Commander (*left center*) Robert F. Overmyer, Pilot (*right center*) Dr. Joseph P. Allen, Mission Specialist (*left*) Dr. William B. Lenoir, Mission Specialist (*right*)
Launch Prep:	Orbiter Processing Facility: 57 days Vehicle Assembly Building: 11 days Launch Pad: 52 days
Launch from KSC:	November 11, 1982; 7:19 A.M. (EST) Lifted off on-time with no delays in schedule.
Mission Duration:	5 days, 2 hours, 14 minutes, 25 seconds **Orbits:** 81
Distance Traveled:	2 million miles
Landing:	November 16, 1982; 6:33 A.M. (PST) Edwards Air Force Base, California Landed on concrete Runway 22.
Wheels-Down to Stop:	9,553 feet
Returned to KSC:	November 22, 1982
Mission:	First deployment of two commercial communications satellites, Anik C-3 for Telesat Canada, and SBS-3 for Satellite Business Systems. First crew of four on an American spacecraft, and first use of mission specialists. Three student experiments and flight data instrumentation packages flown. First spacewalk of shuttle program cancelled due to spacesuit malfunctions.

First Operational Flight

"WE DELIVER" BECAME THE SLOGAN FOR THE FIFTH SHUTTLE FLIGHT, WHICH TOOK PLACE November 11—16, 1982. After the first four flights, which were considered "test" flights, this one was an "operational" mission. This flight, STS-5, carried a payload and, for the first time, had crewmembers aboard who were called "mission specialists." With a crew of four, Columbia launched two communications satellites—Satellite Business Systems' SBS-3 and Telesat Canada's Anik C-3. Led by Vance Brand, an Apollo-Soyuz veteran, the crew consisted of Robert F. Overmyer, pilot, and Joseph P. Allen and William B. Lenoir, both mission specialists.

In addition to launching the two communications satellites, STS-5 conducted a variety of scientific and engineering experiments. A planned spacewalk for STS-5 was cancelled because of a faulty ventilation fan motor in astronaut Allen's spacesuit backpack and a balky pressure regulator in astronaut Lenoir's.

SBS-3

Satellite Business Systems' SBS-3 was the first commercial satellite to be launched from the shuttle. It took place on November 11, 1982.

SBS-3 was designed to provide all-digital communications and was 7 feet in diameter and 21 feet tall when put into orbit. The exterior surface of the satellite was covered with approximately 14,000 solar cells, which generated 1,000 watts of direct current. An on-board power subsystem with rechargeable batteries powered the satellite's communications subsystem, including 10 operational transponder channels. Three types of commercial services were provided by this satellite:

1. *Communications Network Service,* which consisted of high-capacity private networks for all-digital integrated transmission of voice, data, video, and electronic mail for SBS customers with widely dispersed facilities in the United States.
2. *Message Service I,* a high-quality, economical, long distance service for businesses.
3. *Spare Transponder Service,* an offering of spare satellite capacity for communications firms, broadcasters, and cablecasters.

SBS-3 weighed about 7,225 pounds when ejected from the payload bay. Satellite Business Systems is a partnership of Aetna Life and Casualty, Comsat General, and IBM.

Anik C-3

The satellite service of Telesat Canada is the principal means of providing modern voice, message, data, facsimile, and broadcast service to remote and northern parts of Canada. Telesat Canada is owned in equal shares by the government of Canada and a consortium of Canada's telephone companies. The satellite links complemented the terrestrial communications networks and provided a large measure of system diversity to the terrestrial carriers.

The Anik C-3 satellite was the first of a series to provide rooftop-to-rooftop transmission of integrated voice, video, and data communications for Canadian businesses, and to carry newly licensed Canadian pay-TV and other broadcast services. Weighing almost 1,400 pounds, Anik was 7 feet wide and 21 feet tall just like SBS-3. Its solar cells produced more than 1,100 watts of DC electrical power to operate the spacecraft's systems. Anik C-3's antenna coverage included virtually all of populated Canada, with four contiguous spot beams serving the western, west-central, east-central, and eastern regions of the country. Telesat's customers were able to choose regional, half-, or whole-nation coverage, depending on their needs.

Satellite Deployment

Crewmembers Allen and Lenoir deployed SBS-3 from Columbia's payload bay into space at 3:17 P.M. (EST), November 11, and Anik C-3 at 3:24 P.M. (EST), November 12.

The deployments were preceded by about six hours of regularly updated computations of Columbia's orbital parameters—altitude, velocity, inclination, etc. The information was relayed to the private SBS control center in Washington, D.C. and to Telesat Canada's Anik control center in Ottawa. Each of these centers provided NASA's Mission Control Center in Houston with refined payload-insertion information to be passed on to the STS-5 mission specialists, Allen and Lenoir.

Pilot Overmyer then oriented Columbia so that its right wing pointed down toward Earth and its payload bay doors faced opposite to Columbia's direction of movement. Then Lenoir and Allen entered commands into a computer to open the clamshell-like thermal shield of the satellite. This shield protected the satellite from temperature extremes in the open cargo bay. Then they rotated a turntable that spun the satellite and its attached booster rocket (the Payload Assist Module). This spinning action stabilizes the satellite and prevents excessive heating or cooling in space. The Payload Assist Module was designed to propel satellites from low Earth orbit to an elliptical transfer orbit. Finally, explosive bolts were fired to release a clamp holding down a powerful spring. The freed spring then pushed the satellite away from Columbia at a rate of about 3 feet per second.

At this point, control of the satellite shifted to the owner. Brand and Overmyer maneuvered Columbia from a 160-nautical-mile (a nautical mile is 1.1508 statute miles) circular orbit to an oval 160-by-175-nautical-mile orbit, and they rotated Columbia's well-protected belly toward the satellite to protect the orbiter's window from the blast of the Payload Assist Module's solid-fuel rocket motor. When Columbia was about 14 nautical miles above and 17 miles ahead of SBS-3, the satellite was considered clear. The satellite's control center on Earth triggered the Payload Assist Module to blast the satellite into a transfer orbit with a high point of 22,240 miles above Earth. Later, the control center fired the satellite's kick motor to circularize the orbit over the Equator. This was a geostationary orbit, in which the satellite would appear to be standing still if viewed from the surface of the Earth.

Student Experiments Aboard

Three experiments designed by high school students flew aboard the fifth flight. The experiments originated through the Shuttle Student Involvement Project, the joint venture of NASA and the National Science Teachers Association, and had to do with surface tensions of liquids, growth of sponges, and crystals.

Mixing Molten Metals

A metal-mixing experiment aboard STS-5 was the first of 25 Getaway Specials reserved by the Federal Republic of Germany's Ministry of Research and Technology. The main purpose of this experiment was to test a self-operating oven and X-ray unit. It involved periodic X-rays of a molten mixture of mercury and gallium to observe the effects of microgravity on dispersion of mercury droplets into gallium, particle movement due to convection, and other properties. On Earth, molten mercury separates from molten gallium.

Shuttle Engineering Data

STS-5 carried instruments to provide additional information about shuttle aerodynamics, atmospheric-entry heating rates on different sections, the cargo bay environment, and other shuttle properties. One of the experiments studied a strange glow around the tail section of the orbiter. Nighttime photographs taken by the crew on the third shuttle flight revealed an observable glow of unknown origin enveloping certain structural elements of the orbiter, particularly the tail section and engine pods. The glow was not symmetrical and appeared to be more prevalent on the port side of the vehicle. The best guess on the cause of this effect was a recombination of ionospheric oxygen ions.

Landing

Just as on the fourth flight, Columbia had to land on the concrete runway at Edwards rather than on the softer lake bed. Again, recent rains precluded a lake bed touchdown.

During welcome-back ceremonies, Commander Vance Brand talked about the flight. "It's a great day for us. We've been on a fantastic voyage. We're ready to go back again, right now."

Technicians at Cape Kennedy Air Force Station work on the solar panel mechanisms of the two satellites to be carried into low Earth orbit on STS-5. In the background is the Canadian telecommunications satellite, Telesat E, which will be called Anik C-3 in orbit. In the foreground is Satellite Business Systems' SBS-3. Both satellites were propelled from the payload bay into geosynchronous orbit using solid spinning upper stages. The satellite was installed in the orbiter Columbia while it was on the launch pad about one month before liftoff.

First Operational Flight 51

Two members of the four-man STS-5 crew make use of the Johnson Space Center's weightless environment training facility to simulate extravehicular activity for their flight aboard Columbia. Mission Specialist Allen is in the foreground, and Mission Specialist Lenoir is at the top of the photograph. Both men are wearing extravehicular mobility unit spacesuits and are weighted down to achieve neutral buoyancy in the 25-foot-deep pool. The background is a full-scale mockup of the space shuttle's cargo bay area. Scuba divers assist the astronauts in their underwater training session.

Michelle Issel, a student from Wallingford, Connecticut, shows her contribution to STS-5. Issel's experiment dealt with the formation of crystals in a weightless environment. The sponsor for her experiment was Hamilton Standard.

Scott Thomas of Johnstown, Pennsylvania, shows his contribution to STS-5, an experiment to study convection in zero gravity. The experiment was sponsored by Morton Thiokol.

The first two commercial communications satellites scheduled for launch into Earth orbit by a shuttle are pictured in the payload bay of the orbiter Columbia, as seen from the changeout room at Launch Pad 39A. In the first use of the shuttle's ability to carry satellites as cargo, the Satellite Business System SBS-C and Telsat Canada ANIK C-3 rode into space inside the bay, then were ejected and propelled into higher operational orbits. The historic cargo is pictured immediately prior to closing of the bay doors for the last time before the launch of STS-5.

First Operational Flight

The natural beauty of a wildlife refuge graces the foreground of this view of Columbia enroute to Pad 39A.

The STS-5 right solid rocket booster parachutes into the Atlantic Ocean following separation from Columbia. The parachutes remain attached to the boosters until they can be detached by the retrieval crew.

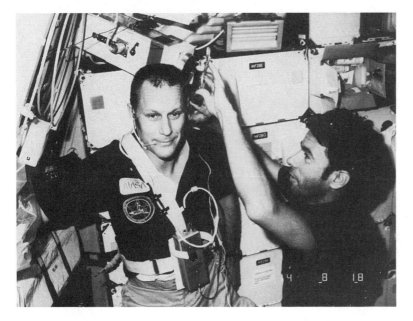

Astronaut Lenoir uses scissors and a brush to trim the sideburns of astronaut Overmyer. An open kit with hygiene supplies inside can be seen attached to one of the stowage locker doors at upper right.

The Satellite Business Systems SBS-3 spacecraft springs from its protective cradle in the cargo bay of Columbia to begin a series of maneuvers that will eventually place it in a geosynchronous orbit. This moment marked a milestone for the STS program—the first deployment of a commercial satellite from an orbiting space vehicle. Both Orbital Maneuvering System pods are seen at center.

A self-exposure of all four STS-5 astronaut crewmembers reveals a bit of their humorous side. The sign held by astronaut Vance D. Brand refers to the successful deployment of two commercial communications satellites on the flight's first two days. Brand is surrounded by (clockwise from top left) astronauts Lenoir, Overmyer, and Allen in the mid-deck area of Columbia.

First Operational Flight

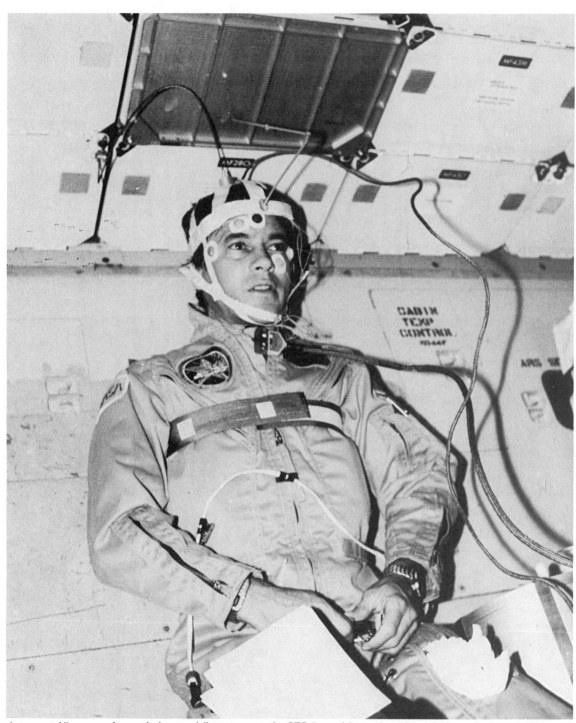

Astronaut Allen, one of two mission specialist astronauts for STS-5, participates in a biomedical test in the mid-deck area of Columbia. A series of electrodes were connected to his face for monitoring his responses to zero gravity. He was assisted in the test by astronaut Lenoir. Dr. Allen is wearing the multi-pieced constant-wear garment for space shuttle astronauts.

Following a successful landing on Runway 22 at Edwards, Columbia is towed to the mate/demate facility. Columbia landed November 16, 1982, following a 5-day, 2-hour mission.

Mission Number: STS-6   **Orbiter:** Challenger

Crew: Paul J. Weitz, Commander (*seated; left*)
Karol J. Bobko, Pilot (*seated; right*)
Donald H. Peterson, Mission Specialist (*standing; left*)
Dr. Story Musgrave, Mission Specialist (*standing; right*)

Launch Prep: Orbiter Processing Facility: 141 days
Vehicle Assembly Building: 7 days
Launch Pad: 126 days

Launch from KSC: April 4, 1983; 1:30 P.M. (EST)
Liftoff was originally set for January 20, 1983. Postponement of more than two months resulted from detection of a hydrogen leak on the Number 1 main engine. An excess hydrogen reading detected in the December 18, 1982 Flight Readiness Firing was confirmed to be a leak in the aft engine compartment by a second Flight Readiness Firing, January 23, 1983. The other main engines were eventually removed to repair fuel-line cracks and were reinstalled. A spare engine replaced the original Number 1 engine. An additional launch delay resulted from contamination of the tracking satellite during a severe storm. The final countdown went as scheduled.

Mission Duration: 5 days, 24 minutes, 32 seconds   **Orbits:** 80

Distance Traveled: 2 million miles

Landing: April 9, 1983; 10:53 A.M. (PST)
Edwards Air Force Base, California (Runway 22)

Wheels-Down to Stop: 7,300 feet

Returned to KSC: April 16, 1983

Mission: First flight of the orbiter Challenger. First Tracking and Data Relay Satellite was deployed on first day of mission. A malfunction of the Inertial Upper Stage unit resulted in placement of the satellite in an improper but stable orbit. Planning for corrective action began immediately. First spacewalk of the shuttle program successfully performed by Peterson and Musgrave. Spacewalk lasted 4 hours, 17 minutes. First use of lightweight external tank and lightweight solid rocket booster casings.

CHALLENGER:
A Second Orbiter

AFTER FIVE SUCCESSFUL FLIGHTS, COLUMBIA STEPPED OFF CENTER STAGE AND MADE WAY for a newcomer, Challenger. When the new orbiter was launched on April 4, 1983, it set a new record for an on-time launch by lifting off the pad only 1/800th of a second after the scheduled takeoff time. Besides being the first flight of Challenger, STS-6 marked the orbiting of the first of a planned quad of Tracking and Data Relay Satellites (TDRSs), which would be essential for shuttle missions to come. The mission also included a spacewalk: the first one by a U.S. astronaut in nine years.

More than 100,000 people welcomed the new spacecraft back to Earth after its maiden space voyage. After completing 80 orbits and gliding to a landing at 1:53 P.M. (EST), April 9, the orbiter needed only 7,300 feet of the paved landing strip at Edwards Air Force Base, coming to a halt with its nosewheel precisely on the centerline of the runway. This precision landing was an example of the skill of Commander Paul J. Weitz, a 50-year-old retired Navy pilot. Weitz was no stranger to spaceflight. He was the pilot of the Spacelab 2 mission, where he logged 672 hours in addition to 2 hours and 11 minutes in extravehicular activity.

New Improved Shuttle

The new orbiter Challenger incorporated many significant advances over its predecessor, Columbia. Challenger's engines were more powerful than Columbia's; this extra thrust gave Challenger an added 4,000 pounds of payload capacity. They were also lighter and stronger, without alterations in basic dimensions.

When launched on its fifth mission in November of 1982, Columbia carried a payload of just over 32,000 pounds. On its first flight into orbit, Challenger lifted an additional 14,500 pounds.

By eliminating stiffeners and brackets to simplify the structure and mountings for the huge main fuel tank, engineers were able to trim more than 10,000 pounds. By redesigning the solid-fuel boosters' engine casings, another 8,000 pounds were eliminated. The new, lighter, recoverable casings were interchangeable with the heavier, but still usable, earlier version. Additional weight reduction measures throughout the spacecraft saved another 2,500 pounds. Ejection seats and their heavy rails disappeared from Challenger's cabin. They were replaced with lighter, operational seats for this mission's four-man crew. The two orbital maneuvering engines at the rear of the orbiter were covered with flexible insulation blankets instead of the individual heat resistant tiles that flew off on the first Columbia mission.

Lightweight honeycomb materials were introduced into the landing-gear doors and in portions of the tail structure. Heat shields for the main engines were also lighter, as were the support structures around engine thrust frames, which were lightened by simplifying their design or combining them with other structural elements.

A variation of the teleprompter appeared for the first time on Challenger's flight deck. Called a Heads Up Display, it presented data and images on an eye-level transparent screen on the flight panel (in front of the commander and pilot, just below the front windows). These images showed velocity numbers, a view of the runway, the ship's angle of descent, and the leveling off for touchdown.

Challenger's first flight had only 22 flight plan anomalies as compared to 82 for Columbia's first flight. One of the more visible ones was torn insulation material on the housing of Challenger's two maneuvering rockets. The damage occurred during liftoff and return from orbit.

Launch Preparation

The launch of April 4 was originally to have taken place 2½ months earlier, on January 20, 1983. A series of gas leaks associated with the orbiter's three high-pressure liquid oxygen and hydrogen engines caused the delay. The leaks had been repaired, and NASA was hoping for an early March launch when severe winds swept through the Cape Canaveral area on the last day of February. These winds were strong enough to blow fine dust particles and other impurities into Challenger's cargo bay where the TDRS was stowed awaiting launch. The satellite and its attached engine, the Inertial Upper Stage (IUS), were contaminated enough to require removal for cleaning and inspection. They were returned to Challenger on March 19. Other engine checkouts, repairs, and verifications continued in preparation for the early April launch, which experienced no delays once the countdown began.

Orbital Operations

Earliest STS-6 plans called for only a two-day shakedown mission for the primary purpose of orbiting the tracking satellite. But the mission was extended to five days to permit the testing of the shuttle spacesuits, which had been plagued with technical problems during the previous mission. A ventilating fan in one suit malfunctioned, and pressure failed to come up to the required levels in the other suit during their initial flight tryouts in November 1982.

In addition to the mission commander, Paul J. Weitz, the other crewmembers were the pilot, Air Force Colonel Karol J. Bobko, 45, and two mission specialists, Dr. Story Musgrave, 47, and Donald K. Peterson, 49. Challenger's maiden flight was carried out by a crew with the oldest average age for an American space crew—48 years and 3 months.

After the spring-actuated release of the tracking satellite, the orbiter backed off and waited for the satellite's booster engine to lift it to its geosynchronous orbit over Brazil. The engine did not fire correctly and resulted in the satellite being placed in the wrong orbit. Instead of being in the plane of the Earth's Equator, the satellite crossed the Equator at a slight angle. The TDRS was the heaviest payload released into orbit by the shuttle, weighing about 5,000 pounds. When placed into orbit its two solar panels extended almost 60 feet from tip to tip (including communications antennae). The Inertial Upper Stage unit, which lifted the satellite from its shuttle orbit of some 200 miles above Earth to a geosynchronous orbit of over 22,000 miles high weighed 32,000 pounds and carried mostly solid fuel propellant.

A NASA/industry team began immediately to fix the problem on the satellite, which was now floating quite uselessly in space. Over the next 59 days they nursed the crippled satellite more than 8,000 miles up toward its proper geosynchronous orbit by judicious firing of the satellite's six thimble-sized thrusters. This achievement is considered to be one of the outstanding engineering feats in space history. At the Goddard Space Flight Center in Greenbelt, Maryland, NASA Administrator James M. Beggs personally triggered the final thruster firing on June 9, 1983 exclaiming, "This puts the shuttle communications program back on track."

In the first U.S. spacewalk in more than nine years, Story Musgrave and Donald Peterson moved out of Challenger's airlock at 4:20 P.M. on April 7. Their activities outside the spacecraft cabin were completed during the 51st, 52nd, and 53rd orbits at 17,500 miles per hour and 177 miles above Earth. They were seen on television during a pass from the Pacific Ocean to the Gulf of Mexico. As they went about their tasks in the cargo bay the astronauts were restrained by 50 foot tethers, which were attached to fixed guide cables along each side of the open compartment. Before they left the airlock, Musgrave and Peterson spent three hours breathing pure oxygen to purge the nitrogen from their blood. This procedure was required to avoid the bends experienced by divers and space travelers when they undergo compression or decompression too rapidly.

Just before the spacewalk began, President Reagan placed a radiotelephone call to the spacecraft. "Please know that all of us, the American people, are proud of your service to your country and what you're doing," he said, and wished the astronauts success. Peterson and Musgrave were the 28th and 29th Americans to walk in space—the first since February 1974 when astronauts in the Skylab crew went outside to retrieve film canisters from some of the orbiting laboratory's optical equipment.

In managing on-time completion of more than 95 percent of their assigned tasks, the astronauts tended three small self-contained Getaway Specials. These included an artificial snow-crystal experiment by a Japanese newspaper, a packaged seed collection flown by a South Carolina seed company, and a five-cubic-foot composite experiment related to metals research, provided by cadets at the Air Force Academy.

Continuing investigations repeating other experiments flown on earlier shuttles included testing of a continuous-flow electrophoresis system, continuing production of small monodisperse latex spheres, and more study of lightning associated with thunderstorms visible below.

Landing

On the day Challenger was to land at Edwards Air Force Base, there was concern about high winds across the desert. John Young, commander of the first shuttle flight, flew up in an observation plane and pronounced, "We're go."

Mission Control then advised the crew that it was okay to land. "All systems are 'go.' A big crowd is waiting for your touchdown."

STS-6 crew in the unofficial "F-Troop" portrait. The four astronauts are wearing the trousers and shirt portions of the space shuttle constant-wear garment outfits, along with some extras, as they take a break in training for their mission. The idea for the costumes spun from the designation of this crew as "F" (the sixth letter of the alphabet) and the tie-in to the well-known TV series "F-Troop."

Astronaut F. Story Musgrave, STS-6 mission specialist, is assisted in a suit donning and doffing exercise in the weightlessness provided by a KC-135 "zero gravity" aircraft. Musgrave and Mission Specialist Donald H. Peterson participated in the exercise as a simulation for their preparations aboard Challenger when they were called upon to perform an extravehicular activity which was postponed from STS-5. The suit is called an extravehicular mobility unit.

The STS-6 crew participates in emergency egress training at Pad 39A as part of the Terminal Countdown Demonstration Test in preparation for the sixth launch of the shuttle.

This view of the Pad 39A launch facility shows the payload canister containing the TRW Tracking and Data Relay Satellite in the rotating Service Structure.

Astronauts Musgrave (at winch device near center) and Peterson set up winch operations at the aft bulkhead as a simulation for a contingency EVA in which they are to run a rope from the winch through a snatch block over to the tilt table for the Inertial Upper Stage. The task would be necessary to return the tilt table to its normal stowed position in the event of failure of an automatic system.

Astronaut Musgrave moves along Challenger's payload bay door hinge line with a bag of latch tools. Dr. Musgrave's task here was to evaluate the techniques required to move along the payload bay's edge with tools. In the lower left foreground are three canisters containing Getaway Special experiments. Part of the starboard wing and Orbital Maneuvering System pod are seen backdropped against the blackness of space.

Deployment of the Tracking and Data Relay Satellite along with its Inertial Upper Stage just after 10:30 P.M. (CST), April 4, 1983.

Astronaut Musgrave participates in a safety tether dynamics checkout procedure during the mission's April 7th extravehicular activity. Dr. Musgrave is tethered to both the near and far slide wire systems.

Challenger makes its first landing shortly before 11 A.M. (PST), April 9, 1983, on Runway 22 at Edwards Air Force Base.

Mission Number:	STS-7 **Orbiter:** Challenger
Crew:	Robert L. Crippen, Commander (*bottom center*) Frederick C. Hauck, Pilot (*bottom right*) John M. Fabian, Mission Specialist (*top left*) Dr. Sally K. Ride, Mission Specialist (*bottom left*) Dr. Norman Thagard, Mission Specialist (*top right*)
Launch Prep:	Orbiter Processing Facility: 34 days Vehicle Assembly Building: 5 days Launch Pad: 24 days
Launch from KSC:	June 18, 1983; 7:33 A.M. (EDT) Liftoff on-time with no delays in schedule.
Mission Duration:	6 days, 2 hours, 24 minutes, 10 seconds **Orbits:** 97
Distance Traveled:	2.2 million miles
Landing:	June 24, 1983; 6:57 A.M. (PDT) Edwards Air Force Base, California (Runway 23)
Wheels-Down to Stop:	8,000 feet (estimate)
Returned to KSC:	June 29, 1983
Mission:	First flight of an American woman into space. Largest flight crew ever launched into orbit aboard a single spacecraft. Continuing validation of Remote Manipulator System through first deployment and retrieval of a spacecraft, the Shuttle Pallet Satellite (SPAS-01). Crew successfully deployed two communications satellites, Anik C-2 for Telesat Canada and Palapa B-1 for Indonesia. Crew also performed proximity operations with the free-flying SPAS-01. Experiments to investigate Space Adaptation Syndrome were carried out. First planned landing at Kennedy Space Center was cancelled due to unacceptable weather at the Florida landing site.

Four Guys and Sally Ride into Space

THE SEVENTH SHUTTLE MISSION WAS UNIQUE. NOT ONLY DID IT CARRY THE FIRST AMERIcan woman into space, but its five-member crew was the largest ever to be launched into space. In addition to Sally Ride, the crew included Robert Crippen, the first repeat shuttle astronaut, Pilot Frederick Hauck, John M. Fabian, and Norman E. Thagard, a medical doctor. The STS-7 six-day mission, from June 18 to June 24, 1983, was the second for the orbiter Challenger.

In the cargo bay were four principal payloads. Two were communications satellites—Anik C-2 and Palapa B-1—both bound for delivery into geosynchronous orbit. The other two returned to Earth with Challenger after testing and experimentation in space. One of these, the Shuttle Pallet Satellite (SPAS-01), was a reusable platform for mounting many types of science instruments and equipment, and was operated during STS-7 both inside the cargo bay and outside as a free-flying spacecraft, another shuttle first. The fourth payload, named OSTA-2 for NASA's Office of Space and Terrestrial Applications, remained fixed inside the bay where four instrument packages ran materials-processing experiments—mixing, melting, or crystallizing such substances as metal alloys and glass in the near weightless vacuum of space. The OSTA-2 experiments were developed by the United States and the Federal Republic of Germany. Also out in the cargo bay, attached to its inside walls, were seven Getaway Special canisters with 22 experiments sponsored by high school and university students, a private company, and two government agencies.

Inside Challenger, two more experiments were carried in the mid-deck cabin. Both of these had flown on previous shuttle missions: the Continuous Flow Electrophoresis System for separating large quantities of pure pharmaceutical fluids in low gravity, and the Monodisperse Latex Reactor, which manufactured small rubber spheres of identical size to be used for precise calibrations and measurements. The total payload weight (both inside and out in the cargo bay) for Challenger's second trip into space was 31,985 pounds.

After a smooth and trouble-free countdown, Challenger was launched on-time from the Kennedy Space Center's Pad 39A at 7:33 A.M. (EDT) on June 18 into a circular orbit at an altitude of 160 nautical miles.

Mission Operations

The first order of business after setting up station in orbit was to release the Anik C-2 communications satellite, sponsored and paid for by Telesat Canada. Approximately 9½ hours into the mission and shortly before the end of their first workday in space, the crew spring-ejected Anik from its spinning platform in the cargo bay and fired Challenger's engines to back the shuttle away from the satellite. A Payload Assist Module rocket motor attached to Anik then fired to begin raising the satellite to its 22,300-mile-high geosynchronous orbit over the equator where it was to be initially used for North America's first direct satellite-to-home pay-TV service.

Challenger's second day in orbit featured a nearly identical deployment of the Indonesian Palapa B-1 satellite, which also was boosted by a Payload Assist rocket to its geosynchronous station over the islands of Indonesia. This was the first in a new series of Indonesian satellites that was to be used for many of that nation's telecommunications needs, including video, telephone, and data transmission. (By the fourth day of the STS-7 mission, both of these communications satellites had reached their desired high orbits, on-time and on-target.)

With the two commercial satellites successfully delivered, the crew turned to other equipment and tasks, including the activation of seven Getaway Special canisters, more than any other shuttle had carried. These varied experiments were designed to test, among other things, the effects of space on radish seeds, germinating snowflowers, liquid mercury, soldering operations, and the social behavior of an ant colony. Two of the canisters featured new Getaway Special technologies, including the first fully automatic experiment (turned on by a barometric switch rather than by the crew) and the first canister with an opening door.

Also on this second day of the mission were the first checkouts of the Ku-band antenna, which was designed to be used for ground communications through the Tracking and Data Relay Satellites in geosynchronous orbit. The antenna's motion and signal acquisition were successfully proven, and it was pronounced ready for operation.

The crew began their third workday with a test that proved that the cabin air pressure could successfully be reduced from 14.7 PSI to 10.2 PSI by controlling the mix of oxygen and nitrogen in the air. This technique was considered an alternative to the 3-hour period of breathing pure oxygen that was required before the astronauts left the controlled environment of the shuttle for outside work. Cabin pressure was dropped while the crew was sleeping to allow a shorter time for purging nitrogen from an astronaut's bloodstream prior to a spacewalk, all to avoid the body's painful reaction to fast changes in atmospheric pressure, commonly known among divers as the bends.

Shuttle Pallet Satellite

During this 30-hour test, the crew also lowered Challenger's orbit slightly to begin a run of experiments with the Shuttle Pallet Satellite (SPAS-01) mounted in the cargo bay. SPAS was the first shuttle cargo financed as a private commercial venture by a European company, Messerschmitt-Bolkow-Blohm of Munich. The SPAS concept was to sell space on the platform—a supporting pallet that provided power and computer processing to attached instruments—to several different customers on subsequent flights. The SPAS itself was designed to be used over and over again.

For STS-7, 10 American and European experiments were fixed to the pallet. Several of these operated while SPAS remained in the cargo bay: microgravity experiments with metal alloys, heat pipes, and pneumatic conveyors; a new instrument to control a spacecraft's position by observing the Earth from a remote sensing scanner that was pointed at different kinds of terrain and land/water boundaries; a mass spectrometer for monitoring gases in the cargo bay and around Challenger's jet thrusters; and an experiment for calibrating solar cells.

SPAS was also part of another shuttle "first." It was released to fly free and then retrieved by the manipulator arm and re-stowed in the cargo bay. This was the first time the space shuttle had truly interacted with another spacecraft in orbit and was a key demonstration of its value and its versatility. As astronaut Crippen put it, "We pick up and deliver."

The free-flight tests began by picking up the SPAS (with the arm controlled by Ride and Fabian) from its berth in the cargo bay, then releasing the pallet while holding it up and away from the Challenger. It was the fourth time the Remote Manipulator Arm had flown on a shuttle, but the first time it had released a payload. The crew then fired Challenger's small reaction-control jets to move the orbiter below and ahead of SPAS as both craft circled the Earth. Finally, Challenger maneuvered up to a point 1,000 feet ahead of SPAS, then slowly moved back to the pallet, closed in, and grabbed it with the manipulator arm.

Later in the same day, Challenger was sent on another series of maneuvers around the satellite, followed by another approach and arm grapple. The point of these maneuvers, which took more than nine hours, was to establish that the shuttle could effectively use new techniques and devices for monitoring another spacecraft (e.g., radar, radio contact, and optical sighting) and then approach the payload and safely grapple it.

During these nine hours, the SPAS pallet itself was by no means passive. It proved a steady platform and effective power supply for its instruments, several of which required microgravity conditions outside of Challenger's cargo bay. One of SPAS's tasks during free flight, completed as planned, was to monitor from a distance the gaseous contaminants in and around the shuttle. Another accomplishment was accumulation of data on the effect of orbiter-jet firings on the SPAS spacecraft at close quarters. A third job was to photograph Challenger during the time when the two spacecraft were separated.

The day after the SPAS pallet's free flight and re-berthing in the cargo bay, the pallet was again removed from the bay for a series of tests of the manipulator arm. With SPAS attached to the arm, Challenger's jets were fired in pulses to test the arm's and the vehicle's stability with a 5,000-pound weight on the remote manipulator. This was part of a continuing series of arm/vehicle motion tests that began with lighter objects during the shuttle's early test flights.

Throughout the mission, such payloads as the Getaway Specials, the OSTA-2 materials-processing experiments, and the Monodisperse Latex Reactor had a requirement to be turned on or off at various times as required by the crew. Later in the flight, the experiments operated automatically. Only the Continuous Flow Electrophoresis System required tending by Ride and Fabian at different times during the entire mission, i.e., when they added sample fluids to the system or collected the final separated products. The results from these payload experiments were evaluated on the ground.

Space Sickness

Norman Thagard, a mission specialist astronaut as well as a medical doctor, was added to the STS-7 crew during December of 1982, after the crew was already in training. His job was to conduct medical tests in orbit that might lead to an understanding of the causes of Space Adaptation Syndrome, the space sickness that has affected a number of shuttle astronauts, just as it did those in earlier manned programs. He performed tests on himself and the other crewmembers, measuring fluid motion and pressure increase inside the head, and checking eye movement and visual perception. Researchers believed that one cause of space sickness was the conflict of signals sent by the inner ear's balancing system and the strange, often upside-down, visual world of weightlessness in orbit.

Landing

With all duties and experimental operations completed as planned, the crew of STS-7 prepared for a return to Earth early on the morning of June 24. It had been hoped that Challenger would make the first landing on the 3-mile-long runway at Kennedy Space Center in Florida, because it took time and money to ferry the orbiter back east from Edwards Air Force Base in California, where all but one shuttle mission had landed (that one being STS-3, at White Sands, New Mexico).

Because of cloudy skies and rain in Florida, however, Challenger did not land at Kennedy Space Center but at the backup site on the desert landing strip at Edwards. Lieutenant General James A. Abrahamson, NASA Associate Administrator for Space Flight said, "We had a small contingent of people doing a sun dance in Florida, but it didn't help."

On June 24 the orbiter left its orbit on the 97th revolution, after which it glided back to Earth and landed on-target at 9:57 A.M. (EDT) after a six-day mission. Because the decision to land in California was made only four hours before it happened, only several hundred people witnessed the landing of the seventh shuttle flight. Challenger's Commander Robert Crippen put the unplanned California landing in a positive light. "We can be ready to land at Kennedy Space Center one minute and land at Edwards Air Force Base the next."

STS-7 crewmembers Crippen, Hauck, Ride, and Fabian are shown with their instructors while training for emergency rescue in an MI-13 armored personnel carrier.

The eight women inset here were part of NASA's 1983 astronaut corps. Clockwise from upper left are Mary L. Cleave, Bonnie J. Dunbar, the late Judith A. Resnik, Anna L. Fisher, Kathryn D. Sullivan, Rhea Seddon, Sally K. Ride, and Shannon W. Lucid.

Mission Specialist Sally K. Ride walks away after a training flight in a T-38 aircraft.

High above the Kennedy Space Center and Florida's Atlantic coastline, Challenger heads toward its second Earth-orbital mission. The orbiter, its two solid rocket boosters, and external fuel tank are just a tiny dot atop the rapidly-lengthening plume of smoke. Astronaut John W. Young, piloting the shuttle training aircraft for his traditional weather-monitoring flight, used a 70mm handheld Hasselblad to record this scene. Launch occurred at 7:33 A.M. (EDT), June 18, 1983.

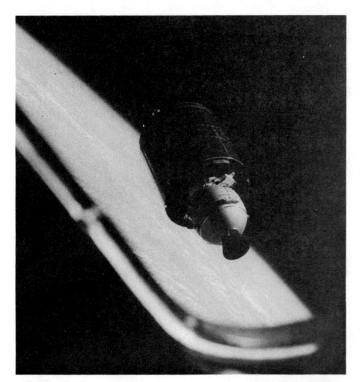

The Canadian Telesat-F (Anik C-2) satellite and the space shuttle Challenger begin their relative separation on June 18, 1983.

The Indonesian Palapa-B communications satellite is just about to clear the vertical stabilizer of Challenger and begin its way toward its orbital destination.

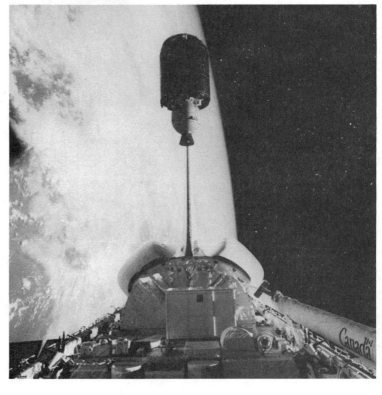

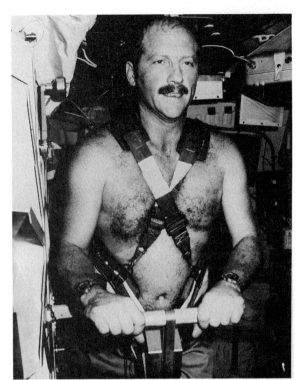

Astronaut Hauck exercises on the treadmill device in the mid-deck of Challenger.

Astronaut Crippen remains at the commander's station when shaving during STS-7.

Four Guys and Sally Ride into Space

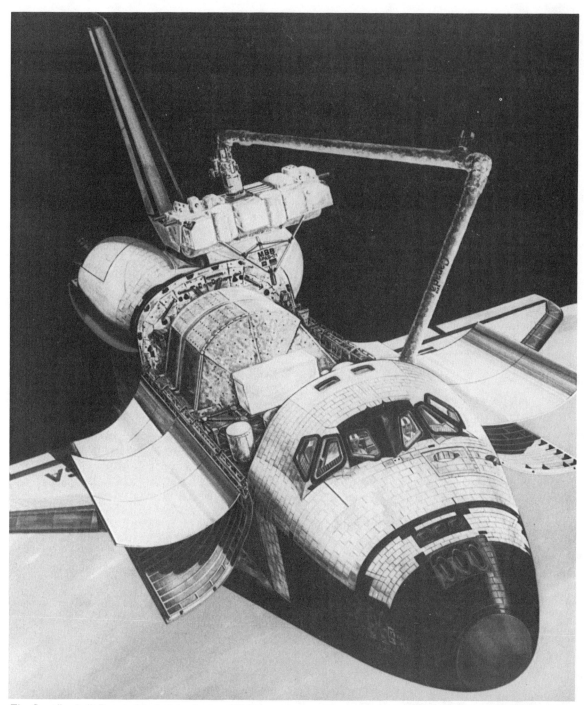

The Canadian-built Remote Manipulator System (RMS), a robot arm-like device, lifts the West German MBB Shuttle Pallet Satellite (SPAS-01) from the cargo bay of Challenger. The satellite was deployed and later retrieved by the RMS during STS-7. This artist's conception provides a perspective that was impossible to capture photographically.

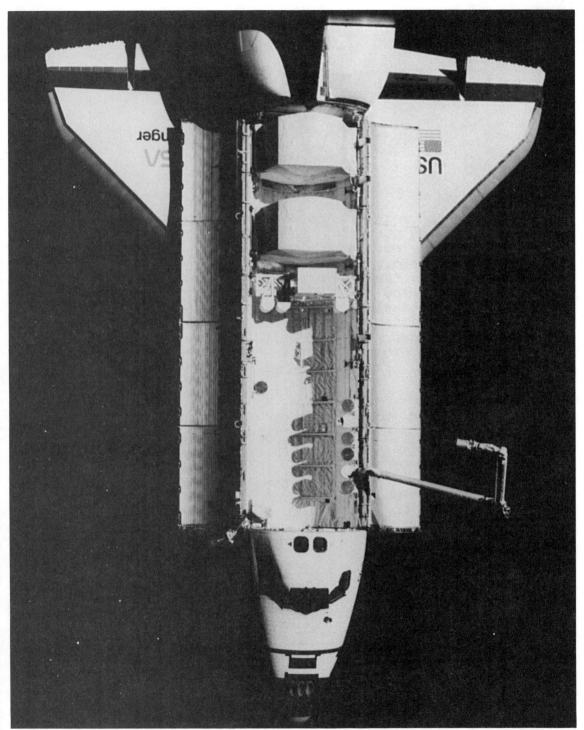
Challenger against the blackness of space was captured by a camera on the temporarily free-flying Shuttle Pallet Satellite.

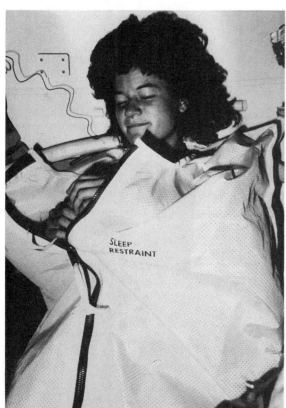

Astronaut Ride was captured at her sleep station in Challenger's mid-deck by a fellow crewmember. Some astronauts choose to sleep in various positions with either their feet or upper bodies anchored. Others elect to use the sleep restraint device, demonstrated here by Dr. Ride.

Challenger's Remote Manipulator Arm grasps the Shuttle Pallet Satellite during proximity operations on June 22. SPAS was returned to a stowed position in the cargo bay for the return to Earth.

An Orbital Maneuvering System engine firing caused this bright glow at the aft end of Challenger on June 18. Also visible are parts of the Shuttle Pallet Satellite, the experiment package for NASA's Office of Space and Terrestrial Applications, the protective cradles for the Indonesian Palapa-B and Telesat Canada Anik C-2 satellites, some Getaway Special canisters and the Canadian-built Remote Manipulator System. The firing took place less than an hour after the deployment of Anik.

Dr. Ride talks to ground controllers from the flight deck of Challenger.

Astronaut Crippen is seen at the commander's station of Challenger as it passes through the Earth's atmosphere on reentry. The friction results in a pinkish glow visible through the forward windows on the flight deck. The flight was scheduled to land at Kennedy Space Center until NASA officials diverted it to Edwards Air Force Base in an eleventh-hour decision. Fog at the Cape reduced visibility at Kennedy's runway to unacceptable levels.

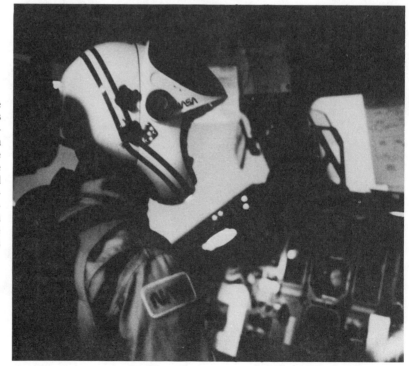

Challenger rests after touchdown on Runway 23 at Edwards.

Mission Number:	STS-8 **Orbiter:** Challenger
Crew:	Richard H. Truly, Commander (*bottom center*) Daniel C. Brandenstein, Pilot (*bottom left*) Dale A. Gardner, Mission Specialist (*top left*) Guion S. Bluford, Jr., Mission Specialist (*bottom right*) Dr. William Thornton, Mission Specialist (*top right*)
Launch Prep:	Orbiter Processing Facility: 26 days Vehicle Assembly Building: 7 days Launch Pad: 28 days
Launch from KSC:	August 30, 1983; 2:32 A.M. (EDT) Launch was delayed 17 minutes due to weather.
Mission Duration:	6 days, 1 hour, 8 minutes, 40 seconds
Distance Traveled:	2.2 million miles
Landing:	September 5, 1983; 12:41 A.M. (PDT) Edwards Air Force Base, California (Runway 22)
Wheels-Down to Stop:	9,200 feet
Returned to KSC:	September 9, 1983
Mission:	First night launch and landing of a space shuttle. First flight of an American black into space. Successful deployment of the Indian National Satellite, INSAT-1B, a multipurpose satellite for India. Payload Flight Test Article was used to test the Remote Manipulator System for large-mass payload capability and to evaluate the elbow, wrist, and shoulder-joint reaction to higher loads. Biofeedback experiments were conducted. Six rats were flown in the Animal Enclosure Module to observe animal reaction in space. Fourth time for Continuous Flow Electrophoresis System to fly, but first time to use live human cells from a pancreas, kidney, and pituitary gland. Other cargo: Development Flight Instrumentation Pallet and Getaway Special canisters. Testing between the Tracking and Data Relay Satellite and Challenger using the Ku-band antenna. Continued investigations of Space Adaptation Syndrome.

Testing a New Communications System

CHALLENGER'S THIRD TRIP INTO SPACE BEGAN AND ENDED IN THE MIDDLE OF THE NIGHT. At 2:32 A.M. (EDT), Richard H. Truly and his crew—Daniel C. Brandenstein, Dale A. Gardner, Guion S. Bluford, Jr., and Dr. William Thornton—began their six-day space flight. Guion Bluford was America's first black astronaut to fly, while Thornton, at 54, was the oldest astronaut to fly.

On this flight, the crew performed the first tests of shuttle-to-ground communications using a new Tracking and Data Relay Satellite (TDRS), delivered India's INSAT-1B satellite successfully into orbit, and exercised the Remote Manipulator System (RMS) arm with its heaviest load to date. It also further developed the experimental Continuous Flow Electrophoresis System, flight tested an incubator and an animal enclosure for use in biological experiments on future flights, carried 260,000 special postal covers into orbit, performed tests on the space environment's effect on astronauts and on man-made materials, and successfully demonstrated both a night launch and a night landing.

Challenger was launched into a 160-nautical-mile orbit from the Kennedy Space Center in Florida at 2:32 A.M. (EDT) on August 30. The launch time was dictated by tracking requirements for the INSAT satellite and also showed that the shuttle could launch at night, when weather conditions in Florida are often more favorable than during the day. The vehicle's ascent into orbit was normal, as was the performance of the first of a new group of lighter, more powerful, solid rocket boosters.

Mission Operations

"Loss of signal" is a familiar expression to astronauts and ground controllers during a spaceflight. Each time the shuttle passes out of range of ground tracking stations there is a complete break in communications until the orbiter comes within radio "sight" of the next station, sometimes more than halfway around the globe. Although there were 15 such tracking stations located in various countries around the world, the shuttle was only in contact with the ground about 20 percent of its total time in orbit on a typical mission. To improve that situation, and also to handle the greater data-flow requirements of modern spacecraft, NASA began installing the Tracking and Data Relay Satellite (TDRS) system in high geosynchronous orbit. The first TDRS had been launched by Challenger in April 1983.

Shortly after Challenger reached orbit on August 30, technicians began the first of a series of 20 tests with the orbiter and the TDRS that ran throughout the mission. These checkouts, most of which were done from the ground with little or no crew involvement, were to evaluate not only the satellite and orbiter hardware but also electronics and computer software at the ground station at White Sands, New Mexico, where signals from TDRS were received and processed.

The orbiter/TDRS radio link was checked out in two frequencies: the S-band and the more powerful Ku-band. Challenger's S-band antennas were located on the outside surface of the vehicle. Ku-band communications used a steerable dish antenna that was mounted in the cargo bay and operated after the bay doors were opened in space. The higher frequency Ku-band was used for the great flows of digital information required by television signals, high-rate telemetry, and the orbiter's text and graphics system (a kind of teleprinter for astronauts to receive copies of printed material sent up to them during a mission). Both text, graphics, and TV transmissions using the TDRS Ku-band relay were successful during the eighth flight, and flight directors were pleased with the clarity and sharpness of both.

Also out in Challenger's cargo bay were one dozen Getaway Special canisters, the most on any shuttle flight to date. Four of these contained experiments: the Cosmic Ray Upset Experiment (CRUX) to determine how high-energy cosmic rays in space can trip changes in the logic state of a computer memory cell; a snow-crystal growth experiment sponsored by a Japanese newspaper, which did not work properly on STS-6 and was redesigned for this flight; a Contamination Monitor Package, one of several STS-8 experiments dedicated to studying atomic oxygen in the thin atmosphere at shuttle orbital altitudes; and a test of ultraviolet photographic film exposed to the cargo bay environment. Eight more Getaway Special canisters and two boxes mounted on an instrument pallet in the cargo bay contained 260,000 STS-8 postal covers for sale to the public following Challenger's return to Earth.

Manipulator Arm

One of the space shuttle's most important pieces of equipment was the 50-foot arm of the Canadian-built Remote Manipulator System, which continued to be the subject of extensive testing on early shuttle flights.

For STS-8, Challenger carried the dumbell-shaped Payload Flight Test Article (PFTA) into orbit to test the arm with its heaviest load to date. The PFTA was simply a passive chunk of aluminum with a heavy lead ballast that brought its weight to 7,460 pounds. More importantly, the chunk of ballast had different grapple points so that it could simulate even more massive payloads to be lifted on subsequent missions. The arm was designed to maneuver a payload of 65,000 pounds.

On Day Three, Dale Gardner used the arm to lift the PFTA out of the cargo bay and move it around in automatic and manually controlled sequences. The crew saw expected amounts of arm vibration when the test article was held straight up from the cargo bay and the orbiter's jets were fired in short pulses. Similarly, instruments on board Challenger checked for vehicle vibrations caused by the arm's motions.

The next day, a similar series of arm maneuvers was run, but this time the test article was grappled from the other end, which required a different range of arm motions to remove it from its berth in the cargo bay. Just as a chair is harder to lift if you pick it up by a leg rather than by its back, the two grapple points presented different inertias for the arm to overcome in lifting the test article. These tests were important for certifying the arm with larger, more massive payloads on later flights. The test article was also designed with screens to block Gardner's vision of payload attach points when he berthed and unberthed the "dumbell" in the cargo bay. This simulated a large payload that would also restrict an astronaut's vision of berthing operations. Gardner was able to secure the PFTA using only video cameras.

Although a computer software problem at the White Sands ground station caused a temporary loss of TDRS relay on the mission's second day, the problem was solved by the next morning, and satellite checkouts resumed. The tests verified that the Challenger and TDRS could automatically lock onto and track each other's signal in both frequencies. Technicians checked out how radio contact was affected by different orbiter attitudes and maneuvers, including rolling the vehicle and pitching it end-over-end. For one test, Challenger's Ku-band antenna was used to acquire the TDRS signal "manually" after it had been deliberately pointed away from the satellite. An encryption technique for sending coded radio messages was also demonstrated to support future missions with classified data.

By mission's end, all of the TDRS communication modes had been proven. There were still problems to be smoothed out with ground software, but the relay satellite showed its readiness for operational support of the STS-9 Spacelab mission. On STS-9 it would be called on to handle large volumes of science data in the high-rate Ku-band mode.

INSAT-1B

The principal payload on this mission was also the only cargo that Challenger left in space—INSAT-1B, a three-in-one satellite that provided India with telephone and data relay service, TV broadcasting, and weather information. Mission Specialist Bluford performed the spring ejection of INSAT from Challenger's cargo bay early on the morning of the second day in space. An attached Payload Assist Module (PAM-D) rocket then began boosting the satellite to geosynchronous orbit, where it arrived on the mission's fourth day.

Additional Activities

Another early task for Mission Specialists Bluford and Gardner involved the Continuous Flow Electrophoresis System (CFES), making its fourth flight on the space shuttle. Previous CFES tests separated out pure proteins by taking advantage of the lack of convection in microgravity, but this was the first time live cell samples were used. Because the cells—human kidney, rat pituitary, and dog pancreas—were living, it was important to perform the CFES runs early and to verify new procedures for keeping them alive both before and after electrophoresis separation. The crew finished the CFES tests on Day Two and also used a new carry-on incubator for keeping the cells healthy in orbit. Speculation was that this research would lead to such medical applications as transplanting live insulin-producing cells into diabetics.

STS-8 also tried out for the first time a new animal enclosure module, a self-contained cage that provided food, water, air, and light for six live rats during the mission. Crew reports showed the rats adapted well to weightlessness. On following missions the enclosure was occupied by other animals used in biological research.

Two more non-biology experiments were performed during the mission's second day. For more than 14 hours, Challenger's nose was held away from the Sun so that engineers could evaluate how the vehicle's canopy area, with its cockpit-like flight deck, would react to the cold. At about the same time, the crew documented on film the performance of an experimental heat pipe mounted in the cargo bay, which was being evaluated for future use in spacecraft thermal control systems. The pipe had a special tape on its outside that changed colors with temperature.

Oxygen Glow and Space Medicine

On their fourth day in orbit, the crew dropped Challenger's altitude down to 121 nautical miles to continue a series of tests of the thin atomic oxygen in the upper atmosphere and how it affects spacecraft structures and materials. This oxygen has been seen to erode certain materials in low Earth orbit and is thought to be the cause of a luminous glow around the orbiter's surfaces photographed on earlier flights.

The crew photographed the glow at night and reported that it was "easily visible to the naked eye" around the tail and the aft-engine pods. Challenger was pointed nose first in the direction of its orbit, in order to drive oxygen particles into the cargo bay where they could interact with different paints, coatings, films, thermal tiles, insulations, and other substances being evaluated for use in space hardware. Material samples were also placed on the manipulator arm to support these oxygen interaction experiments, the results of which would be studied on the ground.

Inside Challenger, Mission Specialist William Thornton—a medical doctor, astronaut, and leading researcher into the nature of Space Adaptation Syndrome (SAS), or space sickness—performed a series of biomedical experiments on himself and the other crewmembers daily throughout the mission. Measurements of eye-hand coordination, response to exercise, and fluid shift inside the body in weightlessness are especially important in determining the causes of SAS. Thornton also administered a student experiment on biofeedback techniques in weightlessness. This was the second consecutive mission with a crewmember assigned specifically to conduct medical experiments.

First Night Landing

With their work in space completed, on what flight directors called the "cleanest mission yet," the Challenger crew returned to Earth early on Labor Day morning, September 5. Because this was to be the shuttle's first night landing, it was decided, as a safety measure, to land at Edwards Air Force Base in California, where a large, flat, lake bed surrounds the landing strip.

"On this Labor Day, you're an inspiration to all of us," said President Reagan to the astronauts in a welcome-home telephone call. "By taking off and landing at night, you've proven that the shuttle is capable of operating generally and under most any condition," the President pointed out.

Challenger performed a de-orbit engine burn on its 97th revolution around the Earth and reentered the atmosphere. Truly and Brandenstein were aided in their final approach by high-intensity xenon arc lights that illuminated the runway. The orbiter had no landing lights of its own, due to the complications of designing outside lights that could withstand the heat of a reentry.

With the experience of dozens of practice night landings in shuttle training aircraft to guide them, Truly and Brandenstein piloted Challenger down out of the darkness and landed on-time and on-target at Edwards, only 300 feet from their runway aim point. Because of the darkness and the lack of any exterior lights on Challenger, persons on the ground could see it only a few seconds before touchdown. Challenger's wheels stopped rolling on the landing strip at 12:41 A.M. (PDT). Just over six days after it had begun in Florida, the STS-8 mission was over.

In a brief ceremony after the landing, Mission Specialist Dr. William Thornton shared his impressions of the flight with those assembled. "Let me say simply that, once in a while, life treats us better than we deserve to be treated. I must admit I know of no point in life that will ever reach this."

Technicians receive the Indian National Satellite (INSAT), its Payload Assist Module (PAM), and support cradle at the Vertical Processing Facility at KSC on July 19, 1983.

Much of Challenger's payload bay, as it was laden for STS-8, is visible in this photograph taken in the Payload Changeout Room at Complex 39's Pad A. At top is the front end of the bay and the windows in the aft bulkhead of the orbiter's crew compartment. The bridge-like structure near the top is the Development Flight Instrumentation Pallet and houses earth-resources and space-science experiments. Extending from near the center to the bottom of the photograph is the dumbbell-shaped Payload Flight Test Article to be used for extensive tests of the handling capabilities of the Canadian-built Remote Manipulating System (RMS). The RMS—in its stowed position—can be seen along the left side. Also visible are the 12 Getaway Special (GAS) canisters. INSAT-1B is stowed in the extreme aft end of the bay, not visible in this photograph.

A powerful electrical storm created an eerie tapestry of light in the skies near Pad 39A in the hours preceding the night launch of STS-8. Driving rains and the dazzling lightning display ceased after this photograph was taken by a remote camera set up by Sam Walton of United Press International.

Challenger lights up the predawn sky as it lifts away from Pad 39A to begin the STS-8 mission with the first nighttime launch of the shuttle era. Challenger's blastoff came 17 minutes late at 2:32 A.M. after mission managers prolonged the final built-in hold until weather conditions became acceptable.

STS-8's crew displays one of the special cacheted postal covers which flew on their mission. The envelope bears a $9.35 postage stamp, intended primarily for Express Mail. Astronaut Truly holds one of the 260,000 covers. Other crewmembers are, left to right, astronauts Bradenstein, Thornton, Bluford, and Gardner.

Astronauts Bluford and Truly fold their arms and stretch out of a rest session.

Astronaut Bluford checks one of the control knobs on the Continuous Flow Electrophoresis System (CFES) experiment which had been flown on all three of Challenger's space missions thus far. CFES is an on-board mid-deck experiment designed to separate biological materials according to their surface electrical charge as they pass through an electric field. McDonnell Douglas worked with NASA during both the operational phase and test phase of the Space Transportation System program on extensive testing with the experiment. It is comparable in size to a household refrigerator.

The Indian National Satellite (INSAT) is about to clear Challenger. The Payload Flight Test Article displays the U.S. flag in the middle of the cargo bay.

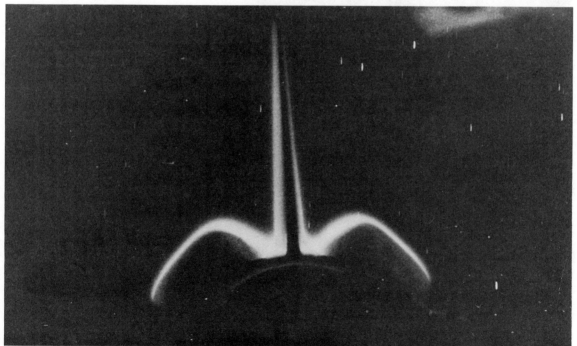

This glowing scene was provided by a long-duration exposure with a 35mm camera aimed at the tail of Challenger. The vehicle was maneuvered to a 120-nautical-mile altitude and flown with the open payload bay to conduct a test titled, "Evaluation of Oxygen Interaction with Materials." Atomic oxygen within the low orbital environment is known to be extremely reactive when in contact with solid surfaces. In the darkened area between the camera and the glowing orbital maneuvering system pods and vertical stabilizer are two trays of test materials.

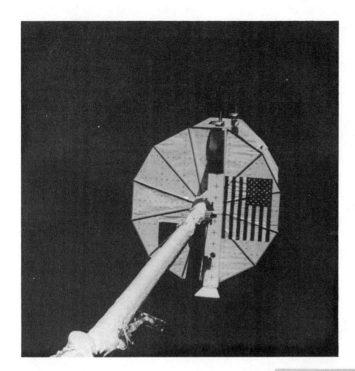

Many hours were spent running tests with the Payload Flight Test Article (PFTA) and the Remote Manipulator System. The bar-bell shaped test device and the arm stand out brilliantly against the darkness of space. The two TV cameras on the "wrist" and "elbow" of the Canadian-built robot arm provided some closeup scenes of the testing.

An astronaut's eye view of two-thirds of the cargo bay, the Earth, and a gibbous moon. The PFTA appears to be awaiting its heavy workout schedule in the middle of the bay.

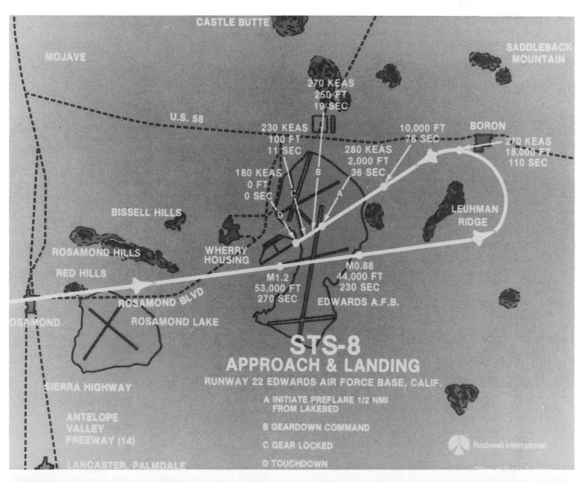

A large number of visitors were on hand at Edwards Air Force Base for the first night landing of the shuttle. Touchdown of Challenger came shortly after 12:40 A.M. (PDT), September 5, 1983.

Testing a New Communications System

Mission Number: STS-9   **Orbiter:** Columbia

Crew: John W. Young, Commander (*bottom right center*)
Brewster H. Shaw, Pilot (*bottom left center*)
Owen Garriott, Mission Specialist (*bottom left*)
Dr. Robert A. Parker, Mission Specialist (*bottom right*)
Dr. Byron K. Lichtenberg, Payload Specialist (*top left*)
Dr. Ulf Merbold, Payload Specialist (European Space Agency) (*top right*)

Launch Prep:

	Flow A	Flow B
Orbiter Processing Facility:	81 days	14 days
Vehicle Assembly Building:	5 days	5 days
Launch Pad:	18 days	21 days

Launch from KSC: November 28, 1983; 11:00 A.M. (EST)
The launch was delayed one month because of a suspect exhaust nozzle in the right-hand solid rocket booster. The problem was discovered after the shuttle had been transported to the launch pad. The shuttle was moved back to the Vehicle Assembly Building and separated from its external tank and solid rocket boosters. The suspect nozzle was then replaced, and the entire shuttle vehicle was restacked.

Mission Duration: 10 days, 7 hours, 47 minutes   **Orbits:** 167

Distance Traveled: 4.3 million miles

Landing: December 8, 1983; 3:47 P.M. (PST)
Edwards Air Force Base, California (Runway 17)

Returned to KSC: December 15, 1983

Mission: First flight of non-astronaut scientists (2) into space, and first foreigner to fly on the shuttle (European Space Agency representative). The flight also marked the first time crewmembers worked around the clock. The European Space Agency and NASA jointly sponsored the Spacelab 1 flight and contributed investigations which demonstrated the capability for advanced research in the areas of atmospheric physics and earth observations, space plasma physics, solar physics and astronomy, and materials science and technology. Space Adaptation Syndrome studies were continued.

Spacelab

SPACELAB, THE EUROPEAN SPACE AGENCY'S SCIENTIFIC WORKSHOP, WAS THE FOCUS OF the ninth space shuttle trip. The ten-day mission, from November 28 to December 8, 1983, was the longest to date. The Spacelab was capable of supporting four scientist-astronauts performing a variety of experiments in biology, physiology, astronomy, and materials science. The Spacelab was being verified on this flight not only for its hardware but also for the entire network of communications between the astronaut-scientists in orbit and scientists on the ground.

With Spacelab in its cargo bay and a six-man crew on board, Columbia was launched from Kennedy Space Center at 11:00 A.M. on November 28. John Young, a veteran of five previous spaceflights, was in command, with Brewster Shaw, Jr. as his pilot. Mission Specialists Owen Garriott and Robert Parker were joined by Byron Lichtenberg and Ulf Merbold, both professional scientists who were assigned to this flight as payload specialists—a new class of crewmembers who are not career astronauts and whose training is more tailored to one particular mission. Merbold, a West German physicist and the first non-American to fly on a U.S. spacecraft, was assigned to Spacelab 1 by the European Space Agency (ESA). This first Spacelab flight was an international venture, with payload weight shared equally by ESA and NASA. Following the launch and settling into a 150-mile-high orbit, the crew opened Columbia's payload bay doors and switched on Spacelab's electrical systems. Just under four hours into the mission, Garriott, Merbold, and Lichtenberg floated from the orbiter mid-deck through a connecting tunnel to enter the shuttle's newest "room" and began initial checkouts.

Among the many firsts for STS-9 were the continuous crew operations in orbit, so that science experiments could operate 24 hours a day. The "Red Team" of Young, Parker, and Merbold rotated on a 12-hour work-then-sleep schedule with the "Blue Team" of Shaw, Garriott, and Lichtenberg. During these 12-hour duty cycles, the commander or pilot would monitor Columbia's systems from the flight deck and change the vehicle's attitude to point instruments toward selected astronomical or Earth targets.

Shortly after entering Spacelab, the Blue Team began the first of the mission's nine full days of scientific investigations. Spacelab 1 was designed to host a wide variety of experiment disciplines in order to demonstrate the lab's versatility. Five major areas—life sciences, materials science, space plasma physics, atmospheric and Earth sciences, and astronomy and solar physics—were represented. Some experiments returned data immediately to scientists in Houston, while other instruments photographed or recorded data to be studied in detail after Columbia's landing.

Life Sciences

Investigation of the human body's reactions to prolonged weightlessness has focused increasingly on two factors: the coordination between vision and the vestibular system (the balance organs of the inner ear), and the shifting of body fluids that occurs in the weightless state. These factors were tested early in the flight, when the crew was first adapting to microgravity, then later in the flight after their bodies and perceptions had adjusted. This data was then compared with tests conducted on the ground before launch and after landing.

The four Spacelab scientist-astronauts took turns being accelerated, spun, and "dropped" with elastic cords (to simulate a fall in microgravity) while being photographed or monitored for physiological reactions. One test investigated the subject's perception of motion by placing his head inside a dome where a pattern of dots rotated to give the sensation of spinning, while a camera recorded eye movements. Another tracked eye motion when the subject was first rotated in a chair, then stopped abruptly. Initial results from these and other Spacelab tests of the visual/vestibular interaction supported the hypothesis that weightless space travelers learn unconsciously to pay more attention to visual and tactile cues about their orientation in space than to cues from their inner ear. In space, the fluid-filled sense organs within the inner ear cease to provide valid information to the brain.

One test, designed by a German researcher, disproved a long-held theory about the workings of the inner ear. The experiment, using Robert Parker and Ulf Merbold as the subjects, showed that convection in the inner ear's liquid is not the only cause of the sensation of spinning that usually results when different densities and temperatures are applied to a subject's ears. Merbold had the sensation even in the weightlessness of space, where there is no such convection caused by temperature and density differences. Another test had all six STS-9 crewmembers trying to distinguish between steel balls of the same size but different mass, to see if their sense of which was heavier had improved during the course of 10 days in orbit. Blood samples were also taken from the crew early and late in the mission to support several experiments concerning the reactions of blood cells, hormones, and the immune system to long-term weightlessness.

Other biology experiments did not use the crew as subjects. An investigation of fungus growth, for example, showed that a fungus culture still followed its own "biological clock" even in a dark environment removed from the usual terrestrial stimuli.

Tests in biology and physiology occupied most of the crew's time during their first two days in orbit.

Materials Science

The manufacturing of new metal alloys and very pure crystals is greatly enhanced in space where there is no thermal convection or mixing of liquids due to gravity. Spacelab 1 carried more than three dozen basic research experiments in materials processing, most of them European. The majority used a Spacelab facility called the Materials Science Double Rack, which had three furnaces for melting or cooling molten metals or crystals. Although two of the furnaces required repairs in orbit, most of the materials samples were processed as planned.

One experiment, sponsored by researchers from the University of Naples in Italy, used the Double Rack's Fluid Physics Module to float a large cylinder of silicon oil between two discs. The crew observed and photographed the fluid's behavior.

On the mission's fifth day, Byron Lichtenberg (who was praised by scientists for saving some of the fluid physics tests that were endangered by balky equipment) became the first person to observe Marangoni Convection—a form of mixing caused by surface tension rather than gravity—in space. Researchers on the ground can only see the phenomenon in tiny samples through microscopes, but the Spacelab experiment allowed it to be photographed on a larger scale. This was an important step toward understanding how fluids experience convection even when gravity is factored out.

Another experiment produced a very homogenous aluminum-zinc alloy, a strong but lightweight metal with its own unique properties. The alloy is difficult to mix on Earth because of the gravity-induced sedimentation of the aluminum during the cooling of the mixture from its molten state. Other tests concentrated on techniques for creating non-metal composites that have the electrical conductivity of metals, and on the study of fluid motion in containers—useful in predicting how rocket propellants slosh around in weightlessness.

One materials-processing experiment was also, in part, a biology experiment. Because crystals can be grown larger in microgravity than on Earth, two pairs of enzyme protein crystals were allowed to grow for a period of 60 hours during the mission. These large crystals allow X-ray studies of protein structure back on Earth.

Space Physics

While life-science and materials-processing experiments were taking place inside the Spacelab module, many of the instruments out on the pallet were aimed toward the Earth and outer space. Several scientists took advantage of the shuttle's orbit through the ionosphere to investigate the surrounding electrically charged environment. To support these space physics studies, STS-9 flew a higher inclination to the Equator—57 degrees—than any other U.S. manned spacecraft, in order to pass through the high-latitude auroral zones.

One instrument, which was designed to generate a powerful beam of electrons to create an artificial aurora in the charged gas surrounding Columbia, did not operate as planned, so the experiment was cancelled. The SEPAC (Space Experiments with Particle Accelerators) instrument did work in its low-power mode, however, and was used with another beam generator on the pallet and with several other instruments to investigate how the Earth's atmosphere and magnetic fields react to controlled artificial stimulation. Natural auroras and lightning were also observed by the imaging cameras used in these experiments, and the electrical interaction of the shuttle and ionosphere was studied.

The failure of one piece of equipment used for space physics studies demonstrated the ability of the orbiter/Spacelab combination to adjust to unforeseen problems and still return scientific data. When a moveable camera for photographing atmospheric light emission failed to lock properly, it was restricted for the rest of the mission to one line of sight overlooking Columbia's wing. Scientists on the ground then requested that the orbiter be moved around to point the cameras in the required directions. John Young and Brewster Shaw complied by adding several new orbiter attitude maneuvers, which saved the experiment.

Earth and Atmospheric Studies

During the early days of the mission, three spectrometers located on the pallet performed highly successful atmospheric-gas analyses in the visible, infrared, and ultraviolet wavelengths. The Spacelab pallet was used as a viewing platform when the cargo bay was pointed at the Earth. Early results were very interesting to experiment sponsors on the ground, who were able to detect such gases as methane, carbon dioxide, and water vapor at higher altitudes than ever before. These findings were important to the study of how the dynamic atmosphere mixes its components.

Inside the module, the crew placed a European Metric Camera in Spacelab's high-quality optical window to take high-resolution mapping photographs of Europe on very wide film. The ESA, sponsor of the experiment, was particularly pleased with the number of photographs taken and with Robert Parker's repair of a jammed film magazine which saved the experiment. As a reward, ground scientists donated some 80 frames of film for the crew to use for their own Earth photography.

A microwave remote-sensing dish, an automatically operated ESA experiment, was not able to do active radar imaging because of an equipment malfunction, but was able to complete studies of the Earth's radar brightness.

Astronomy and Solar Physics

On the mission's third day, the first of Spacelab's three astronomical instruments began to collect data. The Far Ultraviolet Space Telescope, an American instrument mounted on the pallet, took advantage of Spacelab's position above the atmosphere to take ultraviolet photographs of selected galaxies and quasars. The film was returned to the ground for later analysis.

Another instrument, also on the pallet, did studies in the relatively new field of X-ray spectroscopy by surveying the sky's X-ray background and taking spectra for many galactic and stellar sources. In one observation, the X-ray spectroscope detected the presence of iron in the material being drawn into what is believed to be a black hole in the constellation Cygnus.

On the mission's sixth day a third astronomy instrument, the Very Wide Field Camera, was installed by the crew in Spacelab's Scientific Airlock—a chamber for exposing instruments directly to the space environment outside. The camera took ultraviolet photographs to aid in structural studies of the galaxy.

The STS-9 orbit was planned to take Columbia into continuous sunlight, beginning about the mission's seventh day, in order to support several experiments in solar observation. Although the vehicle had to be rotated out of direct sunlight at times because of equipment overheating problems on the pallet, all three solar physics experiments were still able to achieve their objective—precise measurements of the solar constant (the total amount of the Sun's radiant energy that reaches Earth).

Extra Day Added

On Day 6, project managers made the decision to extend the flight by one day to allow more time for experiments. The crew used this extra day to conduct additional experiments in all five disciplines, including several fluid physics demonstrations improvised on the spot by ground scientists and the Spacelab crew.

On their ninth day in orbit, the crew completed and stowed the last of the science experiments, deactivated the Spacelab, and prepared to return to Earth. Just under four hours before the scheduled de-orbit engine burn, however, a jolt from Columbia's attitude-control rockets caused the failure of one of the orbiter's five general-purpose computers. Then, another computer failed. Six hours later, one of the three Inertial Measurement Units used for vehicle navigation also failed.

Although one of the computers was restored to service, the landing was delayed while engineers on the ground evaluated the problem. "We need time to better understand the problem before we commit to reentry," Mission Control told the crew.

To land, Columbia needed only one of the five computers, but according to Lieutenant General James Abrahamson, the NASA shuttle chief, there was concern that "this was a kind of problem that would ripple through all" the computers and systems.

At 3:47 P.M. (PST) on December 8, seven and a half hours after the scheduled landing time, Columbia landed safely at Edwards Air Force Base. One of the computers failed again on landing, and several small fires caused by fuel leaks were discovered around two of the spacecraft's hydraulic power units after touchdown. Although these problems warranted further investigation by project managers, they did not endanger either the Spacelab or the crew.

This fisheye view of Spacelab 1 taken in the Operations and Checkout Building shows payload specialists and technicians in the flight module as they continue performance tests in preparation for the STS-9 launch.

Technicians prepare the Spacelab 1 module and attached pallet for transfer to the Orbiter Processing Facility (OPF) from the O & C Building. The cargo was loaded and sealed into the environmentally controlled payload canister and transported to the OPF for installation into Columbia on August 16, 1983.

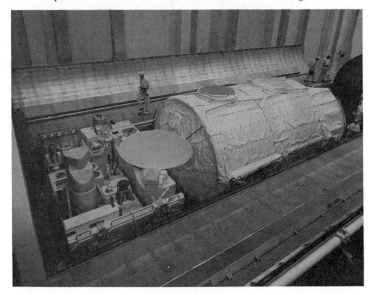

The six-member crew, largest ever for a U.S. spacecraft, lifted off on schedule on November 28, 1983.

Spacelab 103

Astronaut Young takes notes in the commander's station on the flight deck of Columbia.

Astronaut Shaw displays an almost legible printout from the on-board teleprinter.

Astronauts Garriott (left) and Merbold take a break from monitoring experiments aboard Spacelab. Dr. Garriott, one of two NASA mission specialists for this 10-day flight of Columbia, holds in his left hand a data/log book for the solar spectrum experiment, designed to measure energy output in the ultraviolet-to-infrared range. The actual experiment was deployed on the experiment pallet in the open cargo bay of Columbia. Dr. Merbold, a physicist from the Max-Planck Institute and one of two payload specialists for Spacelab 1, holds in his right hand a map for the monitoring of the ground objectives of the metric camera.

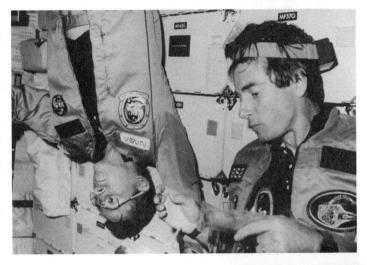

Astronauts Young and Merbold during a meal in the mid-deck of Columbia. Merbold's headband is part of a test to monitor the payload specialists during their waking hours.

Though STS-9 was Columbia's sixth spaceflight, it was the first opportunity for an on-board galley, some of the results of which are shown in this scene on the flight deck. The metal tray makes for easy preparation and serving of in-space meals for crewmembers. This crewmember is seated at the pilot's station on the flight deck. The actual galley is located in the mid-deck.

Mission Number:	STS-41-B (10th Flight) **Orbiter:** Challenger
Crew:	Vance D. Brand, Commander (*seated; left*) Robert L. Gibson, Pilot (*seated; right*) Bruce McCandless II, Mission Specialist (*standing; right*) Ronald E. McNair, Mission Specialist (*standing; center*) Robert L. Stewart, Mission Specialist (*standing; left*)
Launch Prep:	Orbiter Processing Facility: 67 days Vehicle Assembly Building: 6 days Launch Pad: 22 days
Launch from KSC:	February 3, 1984; 8:00 A.M. (EST) Liftoff was originally set for January 29, but was delayed until February 3 so that Challenger's auxiliary power units could be replaced as a precautionary measure.
Mission Duration:	7 days, 23 hours, 17 minutes **Orbits:** 127
Distance Traveled:	2.8 million miles
Landing:	February 11, 1984; 7:17 A.M. (EST) Kennedy Space Center, Florida (Runway 15) First landing of a spacecraft at its launch site. First landing at KSC.
Wheels-Down to Stop:	10,700 feet
Mission:	First untethered spacewalks were performed by astronauts McCandless and Stewart. First in-space use of Manned Maneuvering Unit. West German-built Shuttle Pallet Satellite (SPAS), originally flown on STS-7, became first satellite to be refurbished and flown again. First in-space use of robot arm's Manipulator Foot Restraint. Westar VI and Palapa B-2 satellites were successfully deployed, but failure of Payload Assist Module rocket motors left them in radical low Earth orbits.

Untethered Spacewalks

USING JET-PROPELLED BACKPACKS CALLED MANNED MANEUVERING UNITS (MMUs), TWO astronauts from the crew of the space shuttle's tenth flight flew independently of the mother spacecraft for the first time. The spacewalks were part of preparations for satellite repair missions to come. Also for the first time, the shuttle landed at its Florida landing site. Despite several equipment failures associated with hardware launched from the cargo bay, most of the mission's objectives were accomplished.

For the tenth flight, NASA abandoned the original numbering system used to designate the first nine flights. Instead of "STS-10," the tenth flight was known as "41-B": "4" for the fiscal year (October to October) of the launch, 1984; "1" for the launch site at Kennedy Space Center; and "B" for the second scheduled launch of that year.

From February 3-11, 1984, Challenger orbited the Earth from 165 miles high. In command of Challenger was Vance Brand, who also commanded the fifth flight in November 1982. Pilot Robert Gibson was making his first spaceflight, as were Mission Specialists Bruce McCandless, Ronald McNair, and Robert Stewart. The principal payloads for the 41-B mission were two communications satellites: Western Union's Westar VI and the Republic of Indonesia's Palapa B-2. Also loaded in the cargo bay was the West German-built Shuttle Pallet Satellite (SPAS), the platform loaded with scientific instruments that flew on the STS-7 mission seven months before and was released to fly free and then retrieved by Challenger. On the tenth mission it was used to practice docking maneuvers during the crew's spacewalks. The first satellite ever to be refurbished and flown again, SPAS also carried scientific experiments for West Germany's Ministry of Research and Technology and the European Space Agency.

A two-foot-wide Mylar balloon stowed inside a canister was also carried in the cargo bay for use in practicing rendezvous techniques in orbit. In addition, there were two stowage assemblies with tools and equipment, two MMU backpacks, and a 35mm camera system used to photograph the spacewalks and other cargo bay activities during the mission. This project called "Cinema 360" was a joint endeavor of NASA and a group of planetariums. The special film was to be used in a motion picture about the shuttle program.

Getaway Specials on this trip included experiments in physics, biology, technology, and materials science. They were sponsored by GTE Laboratories, the U.S. Air Force, NASA's Goddard Space Flight Center, and high school and university students from Utah and the University of Aberdeen in Scotland. The experiments were turned on or off by the crew at various times during the mission and operated automatically.

Inside Challenger's pressurized cabin were several more experiments and cargos, including a second "Cinema 360" film camera used by the crew throughout the mission to document typical activities on board the shuttle. Each of these cameras was equipped with a wide-angle lens that covered a half-dome field of view suitable for showing at planetariums.

The Monodisperse Latex Reactor experiment, the mini-factory for making microscopic latex spheres, was flown again on Challenger's mid-deck. The crew turned on the reactor approximately one day into the mission, and after a run of 20 hours it turned itself off automatically.

Another cargo experiment carried inside the vehicle involved electrophoresis, the new technique for separation of protein fluids by their diverse electric charges. For the second time in the shuttle program, live rats were carried on board in an Animal Enclosure Module (life support cage). On this flight, six rats were the subject of a student experiment to study whether or not weightlessness relieves the symptoms of arthritis. Two more experiments, one for monitoring radiation levels inside the vehicle and another for processing material samples in an Acoustic Containment Furnace, were also located on the mid-deck.

Errant Satellites

As on several earlier missions, the shuttle carried commercial satellites into space for launching into geosynchronous orbit. The sequence of events for these launches begins when the satellite is set spinning on a turntable in the cargo bay. It is then released by springs to coast away from the orbiter. Forty-five minutes after deployment, the satellite's attached Payload Assist Module solid rocket motor fires to boost it from low orbit into an elliptical transfer orbit of 260 by 22,300 miles. Later, another small rocket motor fires to change the orbit to a circular one of 22,300 miles.

Using this technique, Westar VI—one in a series of satellites used by Western Union to provide video, voice, and data communication to the United States and its territories—was launched on-time and on-target from Challenger at approximately eight hours into the mission. Soon after the deployment, stations on the ground tracked two objects in an egg-shaped orbit; they appeared to be the Westar VI and its PAM rocket motor. This was confirmed when a ground station regained radio contact with the satellite on February 5. It apparently was undamaged, but had failed to reach the proper orbit. Questions as to why Westar had gone astray forced a rescheduling of the deployment of Indonesia's Palapa satellite from Day Two to Day Four. When Palapa was finally launched, however, it also failed to reach geosynchronous altitude, and assumed a low orbit similar to Westar's. In both satellites, the Payload Assist Module rocket motors failed to boost the satellites into proper orbit. Four shuttle missions later, these same satellites were picked up from their useless orbits and brought back to Earth.

Rehearsal

Much of Flight 41-B was devoted to testing new hardware and procedures for Challenger's next mission, a retrieval and repair of NASA's Solar Maximum Mission satellite, which had a failure in its attitude control system less than a year after it was launched in 1980. The plans for rehearsal went awry when a Mylar balloon was torn to bits while it was being inflated. The balloon was to be used as a "pretend" satellite, and the crew was planning to track it and then practice approaching it. Despite the mishap, useful data were still collected. The crew was able to sight and track a large sunlit fragment of the balloon at the unexpectedly distant range of 40,000 feet.

Outside With No Tether

On the mission's fifth day, Bruce McCandless and Robert Stewart became the first people to fly freely in space, detached from their vehicle. During nearly six hours of outside activity, they successthat of any other spacewalkers in history, thanks to a technique that used lowered cabin pressure to purge the blood of nitrogen. Once outside, McCandless immediately got into his MMU, released it from its latches, and flew away from the orbiter to a distance of half a football field, staying in sight of his crewmates who were watching through overhead windows from inside the Challenger. McCandless then checked out the MMU's maneuvering capabilities further by returning to the bay and then moving away to a distance of 320 feet while being tracked by the orbiter's radar.

McCandless and Stewart moved out into the cargo bay to test the MMUs in space after undergoing a shorter oxygen pre-breathing period than any other spacewalkers in history, thanks to a technique that used lowered cabin pressure to purge the blood of nitrogen. Once outside, McCandless immediately got into his MMU, released it from its latches, and flew away from the orbiter to a distance of half a football field, staying in sight of his crewmates who were watching through overhead windows from inside the Challenger. McCandless then checked out the MMU's maneuvering capabilities further by returning to the bay and then moving away to a distance of 320 feet while being tracked by the orbiter's radar.

While McCandless was test-flying the MMU, Stewart unstowed a device that allowed an astronaut to dock with another spacecraft, such as the Solar Maximum satellite. It fit on the arms of the MMU and had a cylindrical end for locking onto grapple fixtures.

Meanwhile, Stewart checked out another piece of hardware, a Manipulator Foot Restraint platform that fit on the end of the long manipulator arm, allowing a crewmember to be moved around the cargo bay while standing on a stable platform, like a telephone line repairman in a "cherry picker." The crew inside Challenger controlled the arm's motion as Stewart did a number of tests to evaluate arm stress and the stability of the platform. McCandless and Stewart then traded equipment. Meanwhile, McCandless used several tools from the stowage assembly box to practice working on a mockup of the Solar Maximum satellite's main electronic box. Having verified that the backpack unit worked as designed and the foot restraint was a stable work site, Stewart and McCandless reentered the orbiter after nearly six hours. More than a third of that time was spent testing the MMU backpack.

Second Spacewalk

The sixth day of the mission was devoted almost entirely to scientific experiments in the cargo bay. These experiments were carried out on the Shuttle Pallet Satellite, the prototype spacecraft that served as a power supply, data processor, and temperature control to a wide variety of instruments at once. The experiments included two materials-processing experiments in Getaway Special canisters fixed to the platform; experiments in the technology of solar cells, heat pipes, and pneumatic conveyors; a spectrometer for analyzing trace gases in and around the orbiter; an evaluation of a new type of spacecraft attitude control, called a Yaw Earth Sensor; and a remote-sensing camera that recorded images over North America, Central America, and other areas as the Challenger passed overhead.

On Day Seven, Stewart and McCandless suited up for their second spacewalk. Further practice for the future satellite rescue mission was the order of the day. An electrical problem in the 50-foot-long manipulator arm curtailed some of the testing. A second MMU was checked out in various flying and attitude maneuvers, with scrutiny of the unit's ability to stop precisely, hold position, and accurately control direction and speed of motion.

A final chore on this spacewalk was performed by Stewart, who tested a tool for transferring fuel from a storage tank to a satellite or from one satellite to another. Stewart used the tool to open a valve in the Stowage Assembly Box, allowing a mixture of freon and dye to transfer out of a container, much as highly toxic hydrazine fuel would flow in an actual satellite refueling. No dye was seen spilling from the valve, and the transfer appeared to go smoothly. The system was to be checked for leaks after the mission's end.

Landing

After more than six hours out in the cargo bay, the two astronauts reentered Challenger and prepared for their last day in orbit, a day of stowing equipment, turning off experiments, and making ready for the shuttle's first landing in Florida. Shuttle flight planners wanted to land shuttles regularly on the Kennedy Space Center's 15,000-foot-long landing strip to save the time and cost of returning the orbiter from the landing site in California. Challenger was scheduled to land in Florida at the end of the STS-7 mission in June 1983, but bad weather forced a California landing.

On February 11, the weather cooperated. On its 27th revolution, Challenger's maneuvering engines were fired over the Indian Ocean to slow the vehicle and drop it out of orbit. An hour later, Commander Vance Brand brought the orbiter to a smooth stop and at 7:17 A.M. (EST), the tenth space shuttle mission ended, only a short distance from where it had begun eight days earlier.

Mission Specialist Robert L. Stewart models the Manned Maneuvering Unit (MMU), a backpack/motor apparatus allowing much greater freedom of movement than that experienced by any previous spacewalkers.

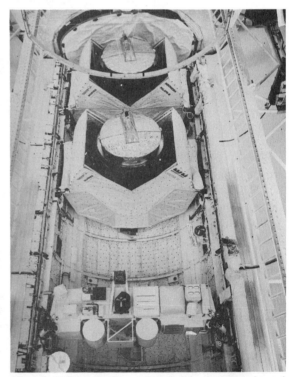

Mission 41-B payloads are installed in the cargo bay of Challenger in preparation for flight. Included are (left to right) Getaway Specials, West Germany's Shuttle Pallet Satellite (SPAS), Western Union's Westar, and the Palapa-B satellite, which was flown for the Indonesian government.

Liftoff of Challenger came at 57.5 milliseconds after 8:00 A.M., February 3, 1984. Under NASA's new numbering system, this 10th space shuttle mission was designated 41-B.

Westar VI and Challenger begin their relative separation after deployment of the communications satellite on February 3.

Palapa-B and the Challenger begin their relative separation after deployment of the communications satellite on February 6.

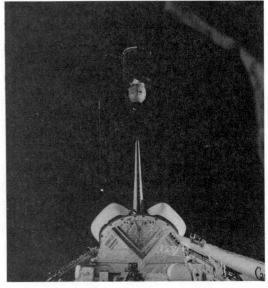

A fixed camera on astronaut McCandless's helmet recorded this rare scene of Challenger some 50-60 meters away during a history-making extravehicular activity, February 7, 1984. Astronaut Stewart, standing beneath the Remote Manipulator System arm, later donned the same Manned Maneuvering Unit that afforded McCandless the freedom of movement to record this image.

Astronaut Stewart tests the MMU on Flight 41-B. (Photo produced by Cinema 360 Inc., Jackson, MS, thru the cooperation of NASA. Filmed by City of Jackson/Davis Planetarium, Jackson, MS. Made possible by a grant from the Gannett Foundation, Rochester, NY. Copyright 1984 Cinema 360, Inc.)

Astronaut McCandless is a few meters away from the cabin of Challenger in this frame. This spacewalk represents the first use of the nitrogen-propelled, hand-controlled Manned Maneuvering Unit.

Astronaut Stewart moves farther away from Challenger's cargo bay during the second of two extravehicular activity (EVA) sessions.

McCandless on February 7 uses the Remote Manipulator System arm and the Mobile Foot Restraint to experiment with a "cherry-picker" concept.

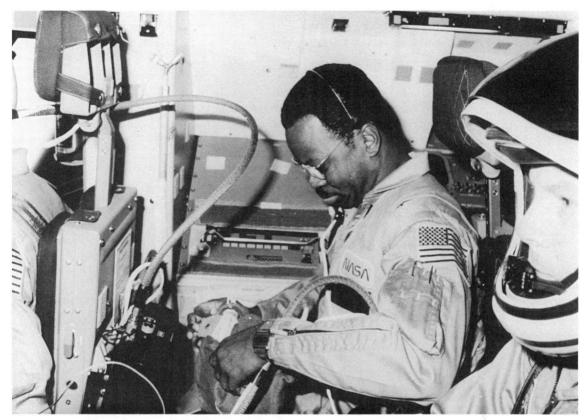
Astronauts McNair (left) and Stewart prepare for the reentry phase of 41-B.

Commander Brand and Pilot Gibson guide Challenger to the first-ever landing of a returning spaceship at Kennedy Space Center. The historic touchdown occurred at 7:17 A.M. (EST), February 11, 1984—about five miles from Launch Pad 39A where the 41-B mission began eight days earlier.

Mission Number: STS-41-C (11th Flight) **Orbiter:** Challenger

Crew: Robert L. Crippen, Commander (*standing; left*)
Francis R. Scobee, Pilot (*standing; right*)
Dr. George D. Nelson, Mission Specialist (*seated; right*)
Dr. James D. van Hoften, Mission Specialist (*seated; center*)
Terry J. Hart, Mission Specialist (*seated; left*)

Launch Prep: Orbiter Processing Facility: 32 days
Vehicle Assembly Building: 4 days
Launch Pad: 19 days

Launch from KSC: April 6, 1984; 8:58 A.M. (EST) **Orbits:** 108

Mission Duration: 6 days, 23 hours, 40 minutes

Distance Traveled: 2.87 million miles

Landing: April 13, 1984; 5:38 A.M. (PST)
Edwards Air Force Base, California (Runway 17)

Returned to KSC: April 18, 1984

Mission: First in-orbit capture, repair, and redeployment of a free-flying spacecraft. First operational use of the Manned Maneuvering Unit, Manipulator Foot Restraint, and extravehicular activity power tools. The attitude-control system and coronagraph/polarimeter electronics box on the "Solar Max" satellite (launched in 1980) were replaced. First time that an orbiter was put into space using "direct insertion" technique. First deployment of Long Duration Exposure Facility, carrying 57 experiments.

Solar Max Repair

THE SPACE SHUTTLE WAS OFTEN COMPARED TO A TRUCK, AND ON THE ELEVENTH FLIGHT IT was a repair truck. The crew of five, led by Commander Robert L. Crippen, repaired and redeployed a satellite while in orbit, another shuttle first. This successful repair mission, 41-C, cost an estimated $48 million compared to the $235 million price for a replacement satellite.

More than four years before, the "Solar Maximum Mission" had been launched by a Delta rocket. It was to serve as an observatory that would study the Sun's flareups and other activity. Early in its development the satellite was dubbed "Solar Max." For the first 10 months, the observatory operated as it was supposed to. Then, it "blew" three hermetically sealed fuses located in a control system that positions the satellite.

Engineers on Earth were forced to figure out a new way of positioning the satellite so that it could continue to at least partially operate. Using on-board equipment in combination with the Earth's magnetic influence, they were able to put the spacecraft in a Sun-pointing position that still permitted three of the seven instruments on board to collect data. Those three instruments were the gamma ray spectrometer, the hard X-ray burst spectrometer, and the active cavity radiometer/polarimeter.

The other four defunct instruments were the coronograph/polarimeter (which the crew replaced on this mission), the ultraviolet spectrometer and polarimeter, the X-ray polychromer, and the hard X-ray spectrometer. The four defunct instruments required pointing accuracy from the satellite and could not function effectively with the spacecraft spinning through space with its longitudinal axis pointed toward the Sun, as it had been doing since the attitude control system failed.

Scenario for Repair

Repairs to be made during the mission included replacing the attitude control system module, replacing the main electronics box on the coronograph/polarimeter, and placing a cover over the gas vent of the X-ray polychrometer. Scientists were hopeful all seven instruments would regain their ability to collect data. However, because of electronics problems that had developed in the hard X-ray imaging spectrometer before the loss of the fuses, scientists felt that there was only a 20-percent chance of obtaining 90-percent use of that instrument.

To do the repair, the crew flew Challenger to the highest altitude above Earth ever achieved by a shuttle. Most missions flew at altitudes of between 150 and 200 miles, but for this mission Crippen and his crew (pilot Francis R. Scobee, and Mission Specialists George Nelson, James D. van Hoften, and Terry J. Hart) rose to 310 miles above Earth and parked about 300 feet away from the ailing Solar Max.

Engineers on Earth then deactivated the attitude control system on Solar Max. The deactivation made it easier to stabilize the spacecraft, but the satellite continued spinning, and the crew still had trouble capturing Solar Max. George Nelson, floating untethered in his Manned Maneuvering Unit, could not latch onto the spinning satellite to slow it down. When he grabbed Solar Max's solar panels, and then fired his MMU's jet thrusters, the satellite ended up in an end-over-end roll. Another attempt to stop the spinning, using the robot arm, also failed.

The next day, the engineers on the ground were able to bring the satellite under control using a secondary attitude control system which allowed the crew to use the robot arm to tuck the satellite away into a cradle in the payload bay. When power cords were attached to the satellite, repairs began. The extensive repair and redeployment of Solar Max came off as planned over a period of several days.

In the cradle, the satellite was tilted forward 25 degrees so its solar arrays could clear the orbiter tail as it was rotated and to provide better access to the attitude control system module. Nelson and van Hoften then positioned themselves to begin replacement of the faulty control system module. To keep from floating away, van Hoften secured his feet on a Manipulator Foot Restraint attached to the end of the orbiter's mechanical arm. Nelson was down below on a portable foot restraint.

Using a special tool, van Hoften unscrewed two retention bolts, removed the faulty control system module, and replaced it with a new module. Then the more difficult task of replacing the main electronics box on the coronograph/polarimeter began. It was especially difficult because the spacesuit gloves were bulky, and the exchange required working with scissors and small screws. During the repair, van Hoften pulled back some thermal protection and installed a hinge so that a panel on the main electronics box could be opened like a door. Using an electric screwdriver, he removed screws that held the panel closed. Challenger was then flown down to 285 nautical miles, to an orbital altitude that was expected to give Solar Max two additional years of effective operations after it was redeployed.

In another spacewalk on another day, Nelson and van Hoften unfolded and taped back the protective thermal blanket, removed the remaining screws on the panel, and secured it open with a special bracket. Van Hoften had to unscrew 22 screws, each with a head no bigger than one-eighth of an inch, while wearing gloves that one journalist compared with boxing gloves. (One astronaut has claimed he can pick up a dime while wearing the gloves—"if given enough time.") The screws held 11 electrical connectors. Van Hoften then had to cut some additional wiring before he could remove the electronics box. The electronics box was removed and stowed in a locker. A replacement unit was then moved into position, at which time Nelson and van Hoften exchanged roles. Nelson re-mated the main electronics box's 11 electrical connectors with clips, eliminating the need to reinstall the previously removed screws. He then removed the panel support bracket, closed the panel door, secured the six panel screws, and reinstalled the thermal protection.

They were joined by Hart who picked up Solar Max on the remote arm and held it off to the side of the shuttle, where engineers on Earth deployed the antenna on Solar Max and conducted some tests with the spacecraft's new attitude control system. The satellite's on-board computer system was reprogrammed completely between spacewalks, with engineers having sent up and checked out 44,000 words of the spacecraft's 48,000 word memory. While outside the shuttle on their spacewalk, the two astronauts were able to witness the ground-controlled movements of the repaired satellite before they returned to the shuttle airlock.

The repaired Solar Max remained on the arm outside the shuttle payload bay throughout the night. The next day Hart positioned the satellite above the shuttle and—after receiving word from the ground that the satellite was okay to release—gently released it from the arm's grasp, placing it back in orbit. The shuttle stationed itself 200 to 300 feet away for approximately two hours and remained in relatively close proximity for a total of eight hours before the astronauts had to start preparations for reentry.

The tentative good news immediately after the deployment was that the satellite worked better than ever. "Indications are we can aim it more accurately than when we put it up," said James Elliot of Goddard Space Flight Center in Maryland.

"The crew rendezvoused with and captured the Solar Max satellite, and they repaired and returned it to orbit," reported Burton Edelson, head of NASA's Office of Space Science and Applications. "It appears to be working well, and that was the purpose of the mission."

Long Duration Exposure Facility (LDEF)

Another purpose of this mission was to release an 8,000-pound platform for 57 scientific applications and technology experiments into Earth orbit for a period of almost one year. The Long Duration Exposure Facility was a 12-sided aluminum structure that was 30 feet long and 14 feet in diameter. The 57 experiments ranged from materials science to medicine to astrophysics. All of them required free-flying exposure in space, but no extensive electrical power, data handling, or attitude control systems. Many of the experiments were relatively simple, and some were completely passive while in orbit. The results of their exposure in space were to be analyzed in laboratory investigations after the platform was returned to Earth. For the academic world, one of the most interesting of the experiments had to do with tomato seeds. 16,000 packets of tomato seeds were aboard the LDEF for later distribution to U.S. classrooms for experiments.

A Beehive of Activity

Flying along with the crew on this mission was a colony of 3,300 honeybees. The purpose was to compare the size, shape, volume, and wall structure of a honeycomb built in space to one built on Earth. Dan Poskevich, a student at Tennessee Technological Institute, Cooksville, Tennessee, devised the experiment. He theorized that by comparing the structures built by a colony of honeybees in zero gravity and normal gravity environments, generalizations could be formed for applications involving other populations of bees, wasps, ants, and related forms.

Honeybees have long been lauded for their selection of a six-sided cylinder as the structural unit that composes the honeycomb. Not only does the hexagon shape hold more honey than a triangular or square one, but it is also strengthened by its contact with adjacent cells.

Two frames were enclosed in a box measuring almost nine square inches. Two cameras photographed each frame continuously during honeycomb construction. To simulate Earth conditions, the environment was regulated using timer-controlled lighting and heating. A food supply was located outside the hive section to supply water, pollen, and nectar. On Earth, a control hive was set up for later comparisons.

Landing

A planned landing in Florida, which was quickly changed to a landing in California, was again the story of the shuttle's return to Earth. It was not the first time, nor the last, that weather at Kennedy Space Center caused the shuttle to land at its second choice location, Edwards Air Force Base in California. In the first eleven flights, there had been three planned landings in Florida but only one took place. The other two were changed to California due to bad weather. Both times Robert L. Crippen was at the helm of the shuttle.

"We do want to have more reliable landings," said Lieutenant General James A. Abrahamson, NASA's shuttle chief. "But we will always be faced with unstable weather. Therefore we're continuing to try to develop, first for safety reasons and later for flexibility reasons, the ability to operate under poor weather conditions."

If the pilot did not have at least seven miles visibility around the landing strip, the shuttle was sent to California or was held off to try again later, as happened in later flights. The California landings played havoc with the regular schedule of flights that NASA was trying to develop. Each time, the orbiter had to be put on top of a 747 jet and ferried back to Florida to be used again.

"It will cost us about five or six days and at least a couple of hundred thousand dollars," said Abrahamson. "The lost turnaround time means the next shuttle flight will be delayed by the same number of days we lose in getting the Challenger back to Florida."

Abrahamson was referring to the fact that Discovery, the orbiter for the 12th flight, could not leave the ground because it was missing some vital engine parts that had been borrowed for use on Challenger. The parts needed to be put back on Discovery.

"The crew was magnificent," said Frank J. Cepollina, head of the space agency's satellite service program. "It was not just business as usual, it was everything all out. We accomplished not only the key objectives of this mission, but the second- and third-order objectives."

The Flight Support Structure (FSS) and the repair equipment for Solar Max are shown here being hoisted from the test stand into the Payload Canister in preparation for transportation to Launch Pad 39A. The FSS was installed into Challenger's payload bay along with the Long Durations Exposure Facility while in the vertical mode at the pad. The FSS was to serve as a cradle for the disabled Solar Max while astronauts attempted to replace faulty equipment.

In a roll of thunder and a pillar of flame that made the morning Florida sun seem pale, NASA's pioneer satellite rescue mission lifts away from Pad 39A. Challenger, carrying a crew of five, began its flawless launch into Earth orbit at 8:58 A.M.

One of the first accomplishments was to place this giant satellite into Earth orbit. Only moments before released from the end effector of the Remote Manipulator System, the Long Duration Exposure Facility (LDEF) is backdropped against ocean waters. The multicolored cylinder carried 50-odd passive scientific experiments representing 194 investigators from around the world. The LDEF program was directed by the Langley Research Center in Hampton, Virginia. The facility was to be retrieved by a later space shuttle crew.

The crew poses for a self-exposure on Challenger's flight deck.

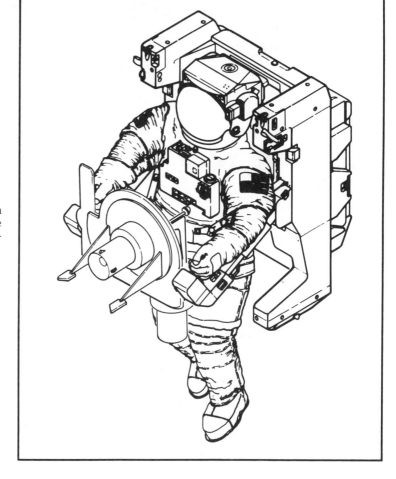

In this sketch, an astronaut holds a trunion pin attachment device, to be used in stabilizing the tumbling Solar Max.

Astronaut Nelson makes the initial excursion to the damaged Solar Max on April 8. Dr. Nelson used the trunion pin attachment device to attempt a "lock-on" with the vehicle, but returned to Challenger after it failed. Days later, the Remote Manipulator System was used to "capture" Solar Max long enough for astronauts Nelson and van Hoften to make the needed repairs.

Solar Max is temporarily docked at Challenger's flight support system for repairs. The RMS arm was used to move the astronauts into position.

Solar Max Repair

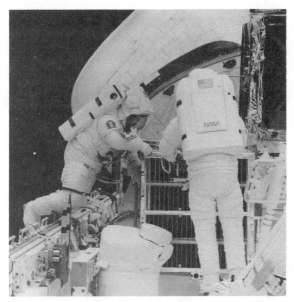

Astronauts van Hoften (right) and Nelson, work cautiously to change a faulty attitude control module on Solar Max. Dr. van Hoften is anchored to a "cherry picker" device which involves a foot restraint/work station connected to the Remote Manipulator System.

Astronaut van Hoften uses a special tool to work on the faulty attitude control module on the captured Solar Max in the aft section of Challenger.

Astronaut Van Hoften finishes off a busy day as satellite repairman.

A view of Hurricane Kamysi in the Indian Ocean, northeast Madagascar.

Astronaut van Hoften holds an aluminum box full of honeybees. The experiment is a "control" duplicate with another colony of the young honeycomb builders on Earth.

Astronaut Hart holds a 70-pound IMAX camera in the mid-deck of Challenger. The 65-mm motion picture camera handles 1,000-ft. rolls which have a running time of three minutes each. Hart, who used a black bag as a darkroom for film changes, commented that the film magazines were much easier to reload in space than on Earth.

Challenger's nose gear touched down at 5:38:22 A.M. (PST) on April 13, 1984.

Solar Max Repair

Mission Number:	STS-41-D (12th Flight)	**Orbiter:**	Discovery

Crew: Henry W. Hartsfield, Commander (*seated; right center*)
Michael L. Coats, Pilot (*seated; right*)
Judith A. Resnik, Mission Specialist (*standing; right*)
Richard M. Mullane, Mission Specialist (*seated; left*)
Steven A. Hawley, Mission Specialist (*seated; left center*)
Charles D. Walker, Payload Specialist (*standing; left*)

		FLOW A	*FLOW B*
Launch Prep:	Orbiter Processing Facility:	124 days	16 days
	Vehicle Assembly Building:	6 days	3 days (ET separation); 7 days
	Launch Pad:	56 days	22 days

Launch from KSC: August 30, 1984; 8:41 A.M. (EDT)
First set for June 25 but scrubbed during T-minus-nine-minute hold due to failure of Discovery's backup general-purpose computer. Attempt on June 26 was aborted at T-minus-four seconds when the computer detected an anomaly in the orbiter's number three engine. Discovery was rolled back to the Vehicle Assembly Building and Orbiter Processing Facility, and the number three engine was replaced. To preserve the launch schedule of future missions, it was decided to remanifest the 41-D cargo to include payload elements from both the 41-D and 41-F flights, and to cancel the 41-F mission. A third attempt on August 29 was delayed when a discrepancy was noted in the flight software of Discovery's Master Events Controller. Discovery's August 30 launch was delayed six minutes when a private aircraft intruded into a restricted area off of Cape Canaveral.

Mission Duration: 6 days, 56 minutes   **Orbits:** 97

Distance Traveled: 2.21 million miles

Landing: September 5, 1984; 6:37 A.M. (PDT)
Edwards Air Force Base, California (Runway 17)
Because the mission was Discovery's first flight, the Edwards desert runway was chosen as primary landing site.

Returned to KSC: September 10, 1984

Mission: First flight of the orbiter Discovery. First deployment of three satellites (Leasat 2, SBS-4, and Telstar 3) on a single mission. First flight of a commercial payload specialist. Heaviest payload carried into orbit (47,500 pounds). First use of a lightweight thermal blanket material on a shuttle's exterior. A 105-foot-tall solar array became the largest structure ever extended from a spacecraft. Continuous Flow Electrophoresis experiment was flown and operated over 100 hours during the mission. IMAX motion picture camera made second of three scheduled flights into space.

DISCOVERY:
A Third Orbiter

DISCOVERY WAS THE THIRD ADDITION TO THE SHUTTLE FLEET AND, AFTER THREE FALSE starts, got off the ground on August 30, 1984.

Originally scheduled for June 25, Discovery's maiden flight was rescheduled for the next day because a backup computer did not work.

The next day's planned launch was stopped with four seconds to go when a computer picked up something wrong in the new orbiter's number three engine. Discovery was rolled back inside the Orbiter Processing Facility, and the suspect engine was replaced.

A third attempt to launch was made on August 29, but launch was again delayed by computer problems.

On the day Discovery finally made it aloft, the launch was delayed by six minutes because a pilot in a small privately owned airplane had illegally entered the restricted airspace of the launch area.

All of the launch delays did result in a certain distinction for the twelfth shuttle flight, however. To keep from ruining the schedule for future missions, NASA decided to skip the next mission and expand Discovery's first payload. This decision gave Discovery's first flight the distinction of carrying the heaviest shuttle payload into orbit, including a record deployment of three satellites from a shuttle cargo bay. Commanded by Henry W. Hartsfield, Jr., Flight 41-D included Charles D. Walker, the first non-astronaut to fly into space under a NASA policy that allowed major shuttle customers to have one of their own people aboard to operate their payloads. Also on this flight was Judith Resnik, America's second woman in space.

Discovery had some improvements over its older sister ships. It was lighter and capable of greater heat loads during reentry than either Challenger or Columbia. The familiar heat-resistant tiles were improved on Discovery, and certain areas of Discovery's aluminum skin were covered with thicker insulation, and graphite-epoxy material replaced some internal aluminum spars and beams in the wings and in the payload bay doors. Discovery's on-board systems were updated and were of more advanced construction than either of its predecessors. In all, the changes resulted in a dry weight of 147,925 pounds for Discovery, compared to a weight of 155,359 pounds for Challenger on the ninth shuttle flight.

Three Satellites

Unlike the tenth shuttle flight (when the satellites were released but failed to reach proper orbit), the three satellites released from the Discovery's cargo bay were successfully deployed and reached their proper geosynchronous orbits.

Eight hours after launch, the first satellite, SBS-4, owned by Satellite Business Systems, was released. It was to be used for television and telephone service.

The second satellite, Leasat 2, was released the second day of the flight. Leasat was owned by Hughes Aircraft Company and leased to the Navy. Leasat was the first satellite designed exclusively for launch aboard the space shuttle. Measuring 14 feet across, the spacecraft was too large to fit in the protective nosecone atop the earlier expendable rockets, such as the Delta or Atlas Centaur.

Unlike the two other satellites carried on the twelfth flight, Leasat was mounted horizontally in the cargo bay. Installation of the spacecraft in the payload bay was accomplished with the aid of a cradle structure. The cradle permitted the spacecraft to be installed lying on its side, with its propulsion system pointing toward the back. Mounted on deployable structures, the antennas could be stowed for launch.

Another unique feature of the Leasat series of satellites was that they did not require a separately procured upper-stage rocket, as had all satellites launched previously. Each Leasat satellite contained its own upper-stage rocket to transfer it from the shuttle's 200-mile-high orbit to a geosynchronous orbit of 22,300 miles above the Earth's equator. Users of the satellite were fixed Earth stations and mobile air, surface, and subsurface stations of the Navy, Marine Corps, Air Force, and Army.

Discovery's third communications satellite was owned by AT&T and was launched the day after Leasat was released. Telstar 3 was the second in a series of AT&T communications satellites representing the latest in technology. It was capable of relaying nearly four times the number of simultaneous telephone calls commonly carried by satellites of the previous generation. In addition, improved batteries and solid-state amplifiers allowed the Telstar 3 series to operate three years longer than the previous generation of satellites—for ten years rather than seven.

After the third satellite was put into orbit, the crew had a call from the Oval Office. "How's it going?" asked President Reagan. "Any surprises?"

Commander Hartsfield replied, "I guess for the five rookies here, it's a big surprise for them. This is such a tremendous ride, you ought to try it sometime yourself."

The President replied that he would like to think that over and asked Judith Resnik, "Is it all you hoped it would be?"

"It certainly is," Resnik replied, "and I couldn't have picked a better crew to fly with."

Secret Stuff

Making its fifth appearance on the space shuttle was the Continuous Flow Electrophoresis System, which was operated for over 100 hours during the mission. Charles D. Walker, an engineer with McDonnell Douglas, operated the "medicine factory" (as it was being called in the press). The system had changed significantly for the twelfth flight, mainly to produce one large sample of several gallons rather than several smaller samples.

McDonnell Douglas had an agreement with Ortho Pharmaceutical Corporation to jointly study the commercial feasibility of using space-based processing to manufacture pharmaceuticals. The pharmaceuticals produced on this mission were company secrets.

Over six gallons of biological material were produced to enable Ortho to conduct research and clinical testing needed to gain Food and Drug Administration approval for "biological substances that we hope to produce for physicians to treat illnesses and maladies on Earth that haven't been treatable before," according to Walker. As a closely held trade secret, the "illnesses and maladies" were unspecified.

During the operation of the experiment, a computer controlled pump ran too fast, and the device shut down twice. Later, Walker operated the experiment by manually running the pump. After the flight, the liquid was put into small vials, frozen, and shipped to the Ortho lab in New Jersey for further processing.

"We loaded it in refrigerated containers and brought it on a company plane to St. Louis," said Susan Flowers, a McDonnell Douglas spokeswoman. "Our biologist met us and took it to the laboratory."

"Everyone is just absolutely tickled to death," Flowers said. "It really has been a tremendous flight. It's taken the efforts of a lot of folks."

Solar Panel

For generating electrical power on future shuttle missions of up to 20 days duration, NASA experimented with a new kind of solar panel on 41-D. This collapsible panel was stretched out and retracted along its 105-foot-long mast to test how it performed in actual use. Panels such as the one tested on this flight, only larger in size, were planned for the NASA space station in the 1990s.

Astronaut Judith Resnik operated the solar panel up and down several times. "We see very little wobbling of the mast as it extends out to 100 feet," said Resnik. "The panels are not sticking the way they did the first time, either, and it looks like it's passing all its dynamic tests very, very nicely."

"I stopped worrying as soon as we got that thing in space," said Gary Turner, the solar-wing program manager for the Lockheed Corporation, which built the $7.6 million panel. "You have no idea how hard it is to test that thing on the ground where it will only support one three-hundredth of its own weight. In space where there is no gravity, it worked just perfectly."

Because the solar panel was tested on this flight primarily to see how easily it could be pulled out, bent back and forth, and then stowed away, not all of the panel was covered with active electricity-generating cells. Only those near the top were active, and the rest were dummies. If fully covered with active cells, the solar panel would produce 12.5 kilowatts of power.

While the panel was extended, the orbiter's steering jets were fired. The mast bent only three or four inches, which was much less than the 10 to 14 inches expected. According to Randy Stone, the flight director, "The array is proving to be very stiff—stiffer than we expected it to be."

Ice

On the fourth day of the flight, a buildup of ice was discovered over a 2½-inch opening on the left side of the shuttle. This opening was where excess water from washing, drinking, and other routine uses was dumped. The weight of the ice chunk was between 8 and 20 pounds and was a concern because it could break off during reentry and damage the orbiter's heat-resistant tiles and aerodynamic control surfaces. The ice had to go, and there were several options: blow high-pressure water or air through the opening, turn the orbiter so that the ice would melt when it faced the Sun, fire rockets to try to shake the ice loose, remove the ice by knocking it off with the 50-foot-long Remote Manipulator Arm, or—the last resort—a spacewalk to knock the ice loose manually.

While the astronauts slept early Monday, the port side of the orbiter was turned toward the Sun, but not enough of the ice melted to do any good. Then Hartsfield fired all of Discovery's maneuvering jets in rapid-fire fashion to rattle the ice loose.

"It didn't do a thing, we've still got the ice blobs." said Pilot Michael Coats. "That gave us a pretty good shake," he said, referring to the force of the jet firings.

In Mission Control, Dick Richards told the crew to begin reducing the air pressure in the cabin in case a spacewalk was necessary.

"Did I hear that right, Dick?" said Commander Hartsfield.

"Yes, sir," came the reply.

Even though a spacewalk was not planned for the mission, two members of the crew, Hawley and Mullane, were trained for just such an emergency spacewalk. Later that day the spacewalk was called off, and Mission Control waited until the next day to solve the problem. The ice chunk was finally knocked off by the Remote Manipulator Arm. It was a delicate move because the ice was located in an area on the shuttle's body where it was impossible to see what was happening at the end of the arm.

While the crew tackled the ice problem, another dilemma cropped up: they could not use the toilet for liquid wastes. On the other side of the orbiter from the "ice blobs" was a similar opening used to dump liquid wastes from the shuttle's toilet. There was a fear that this opening could sprout its own ice chunk, so Mission Control ordered restricted toilet use, with the wastes being stored in a holding tank.

"There is enough water in the toilet's tank for one or two crewmembers to continue to use the john. You can use your imagination and draw your own conclusions about which crewmembers get the use of the toilet," said flight director Randy Stone.

Astronaut Resnik was the only one allowed to use the toilet, and even then, the small holding tank was virtually full when the mission was over. The rest of the crew resorted to a method of liquid-waste removal used during the Apollo era: the plastic bag.

"It's not bad," presumed Dan Germany, shuttle flight equipment manager in Houston. "They haven't complained about it."

Landing

"Now we are back on schedule," NASA informed its satellite customers after the Discovery landed at Edwards Air Force Base early in the morning on September 5, 1984. Despite false starts and some in-flight troubles, the mission accomplished the major portion of its goals and set the pattern for what NASA hoped would be a routine schedule of flights.

In the Vehicle Assembly Building, Discovery is attached to its lifting sling in preparation for rotation to a vertical position and mating with the external tank and solid rocket boosters.

Liftoff of the 41-D mission was scheduled for 8:43 A.M. (EDT), June 26, 1984, and the countdown proceeded flawlessly down to T minus 6.6 seconds, at which the orbiter's three main engines were to be ignited at 120-millisecond intervals. The firing sequence called for ignition in this order: No. 3 (bottom right), No. 2 (bottom left), and No. 1 (top center). Discovery's on-board computers halted the ignition sequence at T minus 4 seconds when the No. 3 engine apparently failed to achieve internal ignition. The No. 2 engine ignited and had built up to 20-percent thrust when it was ordered to shut down after burning for approximately 1.7 seconds. Engine No. 1's start command was overridden by the shutdown order from the on-board computers. Flame, smoke, and steam resulting from the aborted launch attempt are visible in this photograph of the 41-D vehicle on Pad 39A.

DISCOVERY: A Third Orbiter

Discovery lifted off from Pad 39A at Kennedy Space Center for its maiden flight at 8:41 A.M. (EDT), August 30, 1984.

Less than nine hours after launch, the crewmembers deployed the SBS-4 communications satellite. The cylindrical spacecraft spun and rose from its cradle-like protective shield to begin life in space. A number of maneuvers placed it in its desired orbit on August 30, 1984.

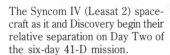

The Syncom IV (Leasat 2) spacecraft as it and Discovery begin their relative separation on Day Two of the six-day 41-D mission.

DISCOVERY: A Third Orbiter 135

OAST-1, the NASA Office of Aeronautics and Space Technology payload for Flight 41-D, represented the largest deployable space structure ever placed in orbit by mankind. In 1975, Lockheed Missiles and Space Co., Inc. developed a demonstration version of a large, flat-folding, flexible-panel solar array. As indicated by this artist's concept, the demonstration model evolved into a flight-ready payload for Discovery's inaugural flight. Objectives of the solar array flight experiment were to demonstrate the readiness of large-area, lightweight, photovoltaic technology; to demonstrate the deployment, retraction, and restowage of the array; and to study the dynamic behavior of a large, flexible, space structure. The structure was extended from a folded stack only three inches thick to a height exceeding that of a 10-story building.

NASA's OAST-1 reached 100-percent extension on the second day of operations—September 2.

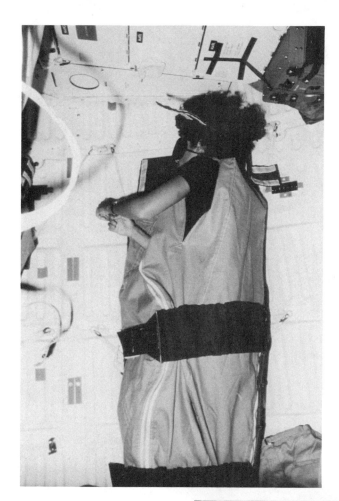

With her sleeping bag held in place by tethers, Mission Specialist Judy Resnik sleeps in weightlessness.

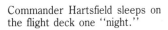

Commander Hartsfield sleeps on the flight deck one "night."

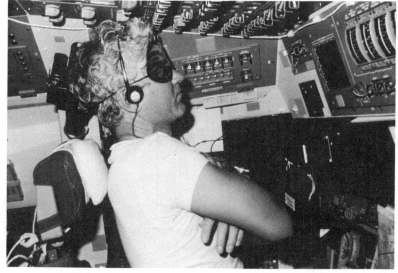

Hartsfield loads a roll of film into the IMAX camera.

Pilot Michael Coats pauses during his meal to monitor spaceflight systems.

This group photo session was one of the lighter moments aboard Discovery.

Astronaut Resnik positions herself on the floor of Discovery's mid-deck to note some items on a clipboard pad. Charles D. Walker, partially out of frame at left, anchors himself with a foot restraint while working at a stowage locker.

Payload Specialist Walker closes a stowage area for biological samples supporting the Continuous Flow Electrophoresis System experiment. The McDonnell Douglas engineer conducted a televised "tour" of the refrigerator-sized processing facility for an audience on Earth. Just beyond Walker's shoulder is the see-through upward-flowing processing canal for the biological samples. An affixed 35-mm camera recorded the flow, during which a separation/purification process occurred. The samples were removed at the top and were later returned to Earth for follow-up studies.

Discovery, about to touch down on Runway 17 at Edwards.

Mission Number: STS-41-G (13th Flight)   **Orbiter:** Challenger

Crew: Robert L. Crippen, Commander (*top; center*)
Jon A. McBride, Pilot (*bottom left*)
David C. Leestma, Mission Specialist
Sally K. Ride, Mission Specialist (*bottom left center*)
Kathryn D. Sullivan, Mission Specialist (*bottom right center*)
Paul Scully-Power, Payload Specialist (*top left*)
Marc Garneau, Payload Specialist (*top right*)

Launch Prep: Orbiter Processing Facility: 69 days
Vehicle Assembly Building: 5 days
Launch Pad: 23 days

Launch from KSC: October 5, 1984; 7:03 A.M. (EDT)

Mission Duration: 8 days, 5 hours, 23 minutes   **Orbits:** 133

Distance Traveled: 4.3 million miles

Landing: October 13, 1984; 12:26 P.M. (EDT)
Kennedy Space Center, Florida

Mission: Largest flight crew ever launched into orbit aboard a single spacecraft. First flight to include two women. Astronaut Kathryn Sullivan became the first American woman to walk in space. Robert Crippen became the first astronaut to fly a fourth shuttle mission. (He was the pilot on his first mission and commander on the other three.) Marc Garneau performed ten experiments dealing with space technology and Earth and space sciences. Garneau also became the first Canadian to fly in space. First demonstration of a satellite refueling technique in space. Earth Radiation Budget Experiment satellite was deployed on Day One, and Shuttle Imaging Radar data was recorded throughout the mission. Oceanographer Paul Scully-Power conducted real-time observations of ocean phenomena from space. Other principal payloads included the Large Format Camera and MAPS, an experiment that measured the amount of air pollution which escapes the MAPS, an Earth's atmosphere and enters space.

Largest Crew Yet

"THE THIRD TIME IS THE CHARM," RADIOED COMMANDER ROBERT CRIPPEN FROM CHALLENger as he approached his first-ever Florida landing after two earlier landings at the second-choice, but more weather-reliable, Edwards Air Force Base in California. Crippen and his crew of four men and two women were completing the 13th shuttle flight. The 41-G mission, from October 5—13, 1983, sported a number of firsts in addition to its record crew of seven:

- first flight to include two women
- first flight of a Canadian payload specialist
- first spacewalk by an American woman
- first demonstration of a satellite refueling technique in space
- first flight with a reentry track that crossed the eastern United States

The landing on this flight took place only 36 days after the last landing, also a record.

Payload

The major payloads being flown on the mission included a package of instruments to record photographic and radar images of the Earth's surface. The images made on this trip would be used for making maps for resource studies and interpretation of geological features.

The Shuttle Imaging Radar-B (SIR-B) produced pictures that looked like black-and-white photos from data collected by transmitting millions of microwave radar pulses sequentially along a broad swath. Characteristics of the terrain alter the signals in different ways—dry sand and rocky surfaces reflect signals differently. The signal returns to the radar, and the unique signature of each reflected pulse is recorded. These millions of radar snapshots are then reconstructed by computers to produce detailed imagery. One of the unique features of imaging radar sensors is their ability to acquire data over virtually any region at any time, regardless of weather or sunlight conditions. The SIR-B antenna consisted of a 35-by-7-foot array of eight panels. The imaging radar first flew on the shuttle during the second flight in November 1981, and an advanced version of the same radar system was on the 13th flight.

The most significant new feature of the imaging radar on its second trip into space was a mechanism that could tilt the antenna to angles of 15 to 60 degrees. This would allow target areas to be viewed from several angles during successive orbital passes. In addition, the antenna could be specially tilted to record surfaces, such as water or vegetation, which show greater detail when viewed from a steep or shallow angle. At a command from the ground, the antenna was programmed to move from one orientation to another at the rate of about one degree per second.

Analysis of data collected over the eastern Sahara Desert on the second shuttle flight showed that the radar signals penetrated deposits of dry, windblown sand and were reflected by underlying geologic formations. The images revealed evidence of ancient river channels in Egypt and Sudan that are believed to be 5-to-40-million years old.

Several experiments in desert regions were prompted by the first imaging results in the Sahara. The imaging radar on this flight was to cover much of the deserts of Egypt and Sudan in search of other geologic formations and evidence of previous human habitation. Other desert areas to be studied were regions in southwest Africa, India, the western coast of Peru, China, central Australia, Saudi Arabia, and the Mojave Desert in California. In addition, the imaging radar on this flight was to be used for groundwater prospecting in Saudi Arabia.

The Swedish island of Oland in the Baltic Sea, a major population center in the Middle Ages, was to be viewed by the imaging radar in an attempt to find ancient Nordic ruins. The island is extremely dry in summer months, and investigators hoped the radar would pass through the thin layer of Earth covering the limestone plateau to reveal any ancient structures or roads.

Other investigations of the radar imaging on this flight were to be studies of tectonic features (East Africa, Canada, midwestern United States, and Turkey), coastal landforms (the Netherlands), meteor impact craters (Canada), vegetation identification (North and South America, New Zealand, Australia, and Europe), tropical deforestation (Brazil), damage to forests by acid rain (Germany), and crop monitoring (Japan, midwestern and western United States, and Australia).

Oceanographic studies were to address the propagation of extreme waves (Agulhas Current, southern Atlantic), internal waves (western and eastern Atlantic), sea ice (South Polar Ocean), icebergs (Labrador coast), and oil spills (Pacific Ocean and North Sea).

In all, the imaging radar on this flight was to collect 50 hours of data covering 18 million square miles of the Earth's surface. However, only a disappointing nine hours of data were collected due to a series of equipment failures. First, the shuttle antenna, which sends data to a communications satellite and then back to Earth, began to move erratically. The motor that points the antenna toward its target satellite could no longer be relied upon to point accurately. Challenger's crew then decided to turn off the steering motor and simply maneuver the entire shuttle to point the antenna in the right direction. To do this, though, they had to stow the eight-panel array. But they could not get it back into its box. It stuck out by one inch, and the container could not be latched. Sally Ride solved the problem by using the remote manipulator to compress the panels enough so that the latch would work.

The shuttle could now be turned to point toward the communications satellite, but there was more trouble. The satellite itself failed for 13 hours. But this time it was caused by the Sun acting up. Sunspots apparently sent out a large enough dose of cosmic rays to blank out the satellite's computer. The Tracking and Data Relay Satellite (TDRS) was designed to dump its computer memory to Earth if it was hit by cosmic rays and needed to be reloaded with new instructions to restore its memory. It took engineers from 9:00 in the morning until 10:00 at night to reprogram the satellite's memory.

"I'm not sure what happened to the satellite, and I'm not sure anyone knows right now," said flight director Cleon Lacefield at the Johnson Space Center in Houston. "All I know is that its memory was wiped out . . ."

All of this moving around—first to move the shuttle so that the radar images could be taken, and then again to send the radar data back to Earth—took up even more valuable time, so fewer hours of data were collected.

Referring to the imaging-radar problems, Dr. Shelby Tilford, NASA director of space science, said, "There is some disappointment that the shuttle's imaging radar has achieved about 40 percent of what we set out to do. In spite of that, we got most of our high priority targets."

The high priority targets included Mount Shasta in the Cascade mountain range in the western United States, huge wheat fields in Illinois, forests in West Germany and Florida, the Imperial Valley in California, and ocean areas off the coast of Chile and in the North Sea where dangerous waves (similar to those off the Cape of Good Hope) form, often without warning.

Other Experiments

Tilford said all the other Challenger experiments worked perfectly. Among them was the Large Format Camera, which took pictures using a 9-by-18-inch negative. Carried on the shuttle for the first time, this camera took photographs (2,300 in all) of every continent on Earth. Weighing in at a hefty 900 pounds, the camera was bigger, more stable, more precise, and more technologically advanced in optics and electronics than its airborne predecessors. From 185 statute miles up, the camera's lens could capture areas as small as 70 feet across (the length of an average single-family house), compared to 270 feet for photos taken by Landsat satellites, the main source of data for Earth resources. A single Large Format Camera picture can cover an area larger than the state of Massachusetts.

Environmentalists concerned about ozone depletion were pleased to see an experiment called the Measurement of Air Pollution from Satellites (MAPS) conducted. Flown previously (November 1981), the MAPS experiment provided information as to what happens to industrial wastes after they enter the atmosphere, determined by measuring the distribution of carbon monoxide in the troposphere on a global scale. The equipment for MAPS consisted of sensors, a digital tape recorder, and an aerial camera. Upon reaching Earth-viewing position, MAPS was turned on, and after a 30-minute warm-up period it began to acquire data.

Another important experiment was the Feature Identification and Location Experiment (FILE), designed to help develop equipment that would make remote-sensing instruments more efficient. This was the first step toward the technology needed to select data at the sensing stage, thus reducing the data load and speeding data dissemination for Earth-observing missions. Except for the absence of a sun sensor, the FILE instrument was the same as on the second shuttle flight. The sensor was eliminated to add greater flexibility for alternative ground commands and operation by the crew.

ERBE

The only satellite released on this flight was the Earth Radiation Budget Experiment (ERBE). Launched on the first day of the mission, its purpose was to accurately measure the amount of solar energy absorbed in different regions of the Earth and the amount of thermal energy emitted back to space.

In the tropics, Earth absorbs more energy than it emits. In the winter polar regions, Earth receives no solar radiation, though it continues to emit radiation. The difference in temperature causes the atmosphere and oceans to circulate and transfer energy, creating Earth's weather and climate. The amount of thermal energy that the Earth reflects back into space, and climate, are changed by several processes: variations in the Sun's output of radiation, veiling of Earth by volcanic dust, increases in atmospheric carbon dioxide from fossil fuel burning, and other processes not yet fully understood.

The satellite's primary goals were to determine, for at least one year: Earth's average monthly regional, zonal, and global radiation budget; seasonal movement of energy from the tropics to the poles; and average daily variation in radiation (computed on a monthly basis) for a 621-square-mile region. Secondary goals were to ensure the accuracy and efficiency of the satellite's instruments and to develop analysis techniques for a highly reliable, economical, and long-term system for climate monitoring and mathematical modeling of climate patterns. The Earth's radiation budget had been studied for several years with instruments on sounding rockets, balloons, and satellites, but the studies provided limited measurements, incomplete coverage, and only sporadic observations. The advent of shuttle operations advanced the state of the art for acquiring this type of data.

The 5,087-pound satellite had three instruments and was lifted out of the orbiter's payload bay by Sally Ride, using the remote manipulator arm.

Refueling

The first spacewalk of an American woman, Kathryn D. Sullivan, took place on this mission when she and David C. Leestma entered the payload bay to practice refueling a satellite. On the three-hour walk, Leestma and Sullivan installed a valve assembly into simulated satellite propulsion plumbing. The simulated plumbing duplicated that of existing Landsat spacecraft, prime contenders for orbital refueling and refurbishment. Landsat and other existing satellites were not designed to be refueled, and the installation of access valves was a prerequisite if such operations were to be undertaken later.

"I love it," said Kathryn Sullivan as she drifted out for her historic walk in space.

Landing

By the time the landing of the 13th shuttle flight took place at Kennedy Space Center, the idea of routinely landing in Florida had almost become a wish. Prior to this mission, the shuttle had landed at the same place it had taken off from only once out of four attempts. For Robert Crippen, who was aboard a shuttle flight for a record fourth time, it was the first time he was not waved off from Florida and sent to the alternative site in California.

"You've outfoxed us again, Crip," astronaut Dick Richards quipped from Mission Control in Houston. "You landed at KSC, but the beer's been sent to Edwards."

A new approach path was tried on this flight. Never before had the shuttle landed by streaking over parts of central North America.

Challenger returned to Earth with only minor damage to some engine parts, which occurred during ascent. "We think we saw a couple of other dings, but the bird looks in reasonably good shape." said Jesse Moore, NASA shuttle chief.

For Canadian Marc Garneau, it was a quiet flight. Throughout the flight he was asked by the Canadian press for his impressions of what it was like being the first Canadian in space. He kept silent. Canadian reporters started referring to him as "The Right Stiff" in a humorous parody of the best-selling book about astronauts called *The Right Stuff*.

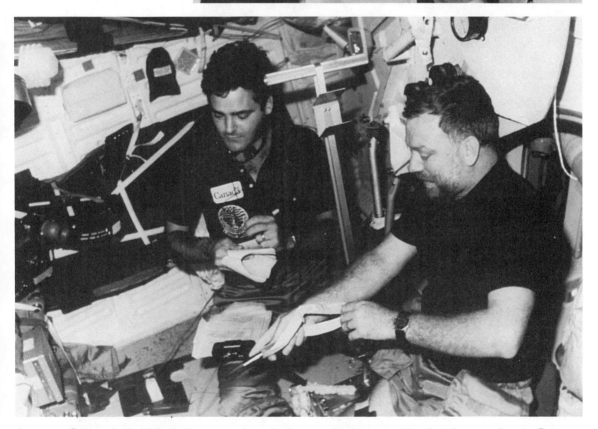

Astronaut Sullivan floats into a middeck scene to join a stationary pair of crewmembers—astronauts Crippen (foreground) and McBride. The white protruding item near the stowage lockers is a Krimsky rule, part of a near-vision acuity experiment.

Astronauts Garneau (left) and Scully-Power are pictured prior to conducting one of the Canadian experiments. Between the two payload specialists is the treadmill exercise device. Scully-Power spent much of his time on the eight-day flight observing and photographing Earth's oceans.

Largest Crew Yet 147

Drs. Sullivan (left) and Ride show off what appears to be a "bag of worms," a product of their creativity. The "bag" is a sleep restraint, and the majority of the "worms" are springs and clips used with the sleep restraint in its normal application. Clamps, a bungee cord, and velcro strips are other recognizable items in the "creation."

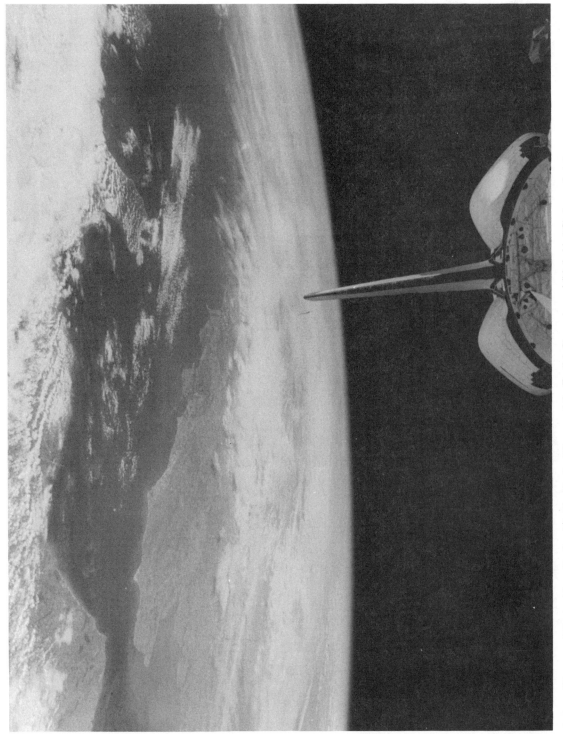

A view of France and England from Earth orbit. The Strait of Dover and the English Channel are visible behind Challenger's tail.

Largest Crew Yet

Challenger makes its approach landing on the 15,000-foot Shuttle Landing Facility at Kennedy.

The touchdown represented the second shuttle landing at the Florida launch site, a practice that NASA wanted to become the rule.

Mission Number: STS-51-A (14th Flight) **Orbiter:** Discovery

Crew: Frederick H. Hauck, Commander (*bottom*)
David M. Walker, Pilot (*left center*)
Anna L. Fisher, Mission Specialist (*right center*)
Dale A. Gardner, Mission Specialist (*left*)
Joseph P. Allen, Mission Specialist (*right*)

Launch Prep: Orbiter Processing Facility: 37 days
Vehicle Assembly Building: 5 days
Launch Pad: 17 days

Launch from KSC: November 8, 1984; 7:15 A.M. (EST)
First attempt on November 7 scrubbed during built-in hold at T minus 20 minutes, due to shear winds in upper atmosphere.

Mission Duration: 7 days, 23 hours, 45 minutes **Orbits:** 127

Distance Traveled: 3.3 million miles

Landing: November 16, 1984; 7:00 A.M. (EST)
Kennedy Space Center, Florida (Runway 33)

Mission: Discovery's second mission in space; the first flight ever to deploy two communications satellites and retrieve two other disabled satellites. On Day 2, Canadian communications satellite Anik D-2 (Telesat H) was deployed into geosynchronous orbit. Wearing Manned Maneuvering Units, propelled by pressurized nitrogen gas, Allen and Gardner retrieved the malfunctioning Palapa B-2 and Westar VI satellites. The two satellites had originally been deployed during Mission 41-B in February 1984. Use of the satellites was lost when the Payload Assist Modules failed to boost them into usable orbit

Satellite Pickup and Delivery

FOR THE 14TH SHUTTLE FLIGHT, THERE WERE TWO SATELLITES IN THE CARGO BAY WHEN the vehicle was launched November 8, 1984. When the shuttle Discovery landed seven days later, it still had two satellites in the cargo bay, but not the same two. For the first time, the shuttle had picked up satellites and returned them to Earth for repair and resale. After successfully launching the Canadian communications satellite Anik D-2 (Telesat H) and the Hughes Leasat 1, Discovery and crew proceeded to a slightly higher orbit where two nearly identical communcations satellites were floating uselessly in space after their booster rockets had failed to take them up to their planned orbits in a February 1984 launch.

The two derelict satellites were owned by their insurers, who had paid off the original owners as a result of failure of the satellites' rockets to push them to their planned orbits of just over 22,300 miles above the Earth. Initially, the group of insurers that owned the Palapa satellite had agreed to pay NASA $4.8 million to rescue the satellite. But two months before the rescue attempt, NASA signed on another group of insurers to rescue the nearly identical Westar 6. The agency would rescue both satellites for $5.5 million.

"We've deployed satellites before, we've picked up satellites before, we've rendezvoused before, and we've repaired a satellite before," said flight director Jay Greene at Johnson Space Center. "But we've never before done all these things together on one flight."

After a one-day delay caused by high winds, Discovery lifted off at 7:15 A.M., November 8, 1984. Commanding the five-member crew was veteran Frederick H. Hauck, with Pilot David M. Walker and Mission Specialists Anna L. Fisher, Dale A. Gardner, and Joseph P. Allen comprising the rest of the crew. Anna Fisher (who became the first mother to fly in space) and David Walker were the only rookies in the crew.

Delivery and Pickup

After rising to the usual orbit of approximately 190 miles above the Earth, the crew launched the Canadian satellite and, the next day, completely emptied the cargo bay by launching the second satellite, Leasat 1. At that time Discovery was approximately 5,000 miles behind the two stray satellites, which shared an orbit 224 miles up, but were separated by 690 miles.

In order to get to Palapa and Westar, Discovery began a complicated series of maneuvers. By firing its steering rockets 44 times, more than in any previous flight, Discovery caught up to Palapa, the first satellite to be picked up. Maneuvering to within a few dozen feet of the satellite, astronauts Gardner and Allen put on their spacewalking backpacks and moved out to tackle the first item of business: to stop the satellite from spinning slowly round and round. They used the "stinger," which was basically a shaft with a wheel-shaped handle at one end. This was inserted up the satellite's empty rocket nozzle to act as something the astronauts could grab onto—the satellites were not built with handles or hooks on them. With a few bursts of the backpacks' thrusters, the spin was stopped. The original plan called for the astronauts to install a metal brace on top of the satellite so that the robotic arm could grab it and place it in the cargo bay. The plan did not work, however, because the brace did not fit. A small protruding "feed horn," used for signal transmission, was sticking out about 1/8 of an inch too far. When the brace was made for this salvage operation the existence of this protrusion was not known and an allowance was not made for it. In the end, human muscle power became the key to getting the satellites into the cargo bay.

In the Palapa retrieval, Allen and Gardner were able to attach the robotic arm to the bottom end of the satellite where the stinger was attached. When they had swung Palapa over to the cargo bay, it was actually upside down and they had to wrestle the satellite into a right-side-up position and lock it into position. So that they would not have to do all this juggling around with the second satellite, Westar, they devised a simpler plan. Allen locked his feet to the end of the robot arm and reached out and clenched an antenna on top of the satellite, while Gardner held on to the bottom. Then, Anna Fisher, operating the arm from inside the cabin, moved the entire ensemble (one 1,200-pound satellite and two men) over to the cargo bay. Westar, already right-side-up, was then locked into the cargo bay, in considerably less time than it took to stow Palapa.

"We have two satellites latched in the cargo bay," announced Frederick Hauck when the job was done.

On the last full day of the mission, the crew tidied up the cabin and completed the one scientific experiment of the mission, growing organic crystals for later study of their optical properties. Also on the last day, the crew held a press conference and chatted with President Reagan about, among other things, making "satellite lifting" a "new high-tech Olympic sport."

On landing at Kennedy Space Center on November 16, 1984, the satellites were examined by technicians and were said to be in good shape, except for a few solar cells that were nicked when the satellites were being placed into the cargo bay.

Discovery in its lifting sling prior to the 51-A mission. The orbiter's nose is made of heat-resistant reinforced carbon-carbon material.

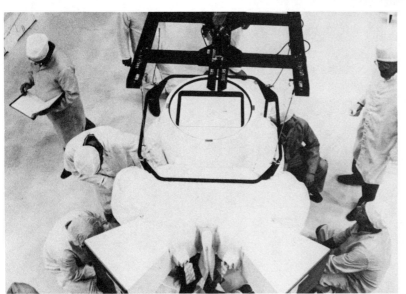

Engineers prepare the Large Format Camera for installation on its Mission Peculiar Experiment Support Structure in the Operations and Checkout Building. The LFC, a high-altitude aerial mapping camera, was flown in the cargo bay of Discovery on its maiden flight. Pictures taken by the camera would be used for qualitative studies in a variety of "earth resources" disciplines, including geology, geography, forestry, and agriculture. The "V-shaped" instrument attached to the camera is the Attitude Reference System, a two-camera array to record star field data at the moment of each LFC exposure. This data would be used to establish positional accuracies for mapping.

Liftoff of 51-A at 7:15 A.M., November 8, 1984.

The busy end effector of the Remote Manipulator System. The hand-like device was at the command of astronaut Fisher during much of the eight-day mission. Dr. Fisher worked long hours on November 12th and 14th to manipulate the arm to aid astronauts Gardner and Allen as they captured two stranded communications satellites. During flight, the wrist television camera (center of frame) recorded many dramatic close-up scenes.

Astronaut Allen holds onto the top portion of the Westar VI satellite while astronaut Gardner (out of frame) works to free a "stinger" device from the opposite end. The large tool, called a stinger apogee kick-motor capture device, allowed Gardner to achieve a hard dock with the previously stranded spinning craft. Dr. Allen is standing on a mobile foot restraint attached to the RMS arm's end effector. The arrangement created a "cherry picker" for the space workers. Dr. Fisher worked the controls from Discovery's aft flight-deck panel during this and earlier satellite-retrieval EVA on this flight.

The five-member crew celebrates a successful mission. The reference to the Eagle has to do with the Discovery crew's mascot, which appears in the official crew portrait and insignia.

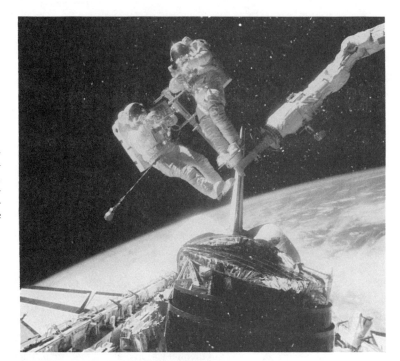

In this view of Westar VI post-retrieval activity, astronaut Gardner (left) holds a For Sale sign, making light reference to the status of the recaptured communications. Astronaut Allen stands on the mobile foot restraint.

Astronaut Allen's view of astronaut Gardner. Allen is reflected in Gardner's helmet visor.

Satellite Pickup and Delivery

Mission Number:	STS-51-C (15th Flight)
Orbiter:	Discovery
Crew:	Thomas K. Mattingly, Commander (*bottom right*) Loren J. Shriver, Pilot (*bottom left*) James F. Buchli, Mission Specialist (*top center*) Ellison S. Onizuka, Mission Specialist (*top right*) Gary E. Payton, Payload Specialist (*top left*)
Launch Prep:	Orbiter Processing Facility: 35 days Vehicle Assembly Building: 14 days Launch Pad: 20 days
Launch from KSC:	January 24, 1985; 2:50 P.M. (EST) First set for January 23, but scrubbed due to freezing weather conditions. Orbiter Challenger was originally scheduled for the mission, but thermal tile problems forced substitution of Discovery.
Mission Duration:	3 days, 1 hour, 33 minutes
Landing:	January 27, 1985; 4:23 P.M. (EST) Kennedy Space Center, Florida (Runway 15)
Wheels-Down to Stop:	7600 feet
Mission:	Space shuttle Discovery's third trip to space. First mission totally dedicated to the Department of Defense. The U.S. Air Force Inertial Upper Stage (IUS) booster rocket was deployed and successfully met its mission objectives.

First Secret Military Mission

THE SHUTTLE WAS NOT THE ONLY THING LAUNCHED ON THE OCCASION OF THE 15TH FLIGHT. An investigation of news leaks about the secret cargo aboard Discovery's third flight was also launched by the government in an effort to find out where leaks were coming from.

All previous shuttle flights had followed NASA's traditional pattern of sharing even the most trivial details about a space mission. Word-for-word communication between the ground and the space vehicle was continuously shared by NASA with anyone who was interested. Thousands of individuals were usually invited to view the shuttle launches and landings.

Not so on the 15th shuttle.

Secrecy Prevails

This mission was the first one dedicated totally to the military, and "mum" was the word. It was the first time in 46 U.S. manned spaceflights that flight information was held back. NASA was uncharacteristically reticent with details of Discovery's third flight—even to the point of not announcing in advance when the launch would occur.

The only detail made public about Discovery's cargo was that an Inertial Upper Stage (IUS) booster rocket "was deployed and successfully met its mission objectives." This category of booster rocket is normally used to boost satellites from the shuttle altitude of a couple of hundred miles to a geosynchronous orbit of 22,300 miles. The same type of rocket booster failed on the sixth shuttle flight when a tracking satellite was being put into orbit. Fortunately, the tracking satellite ended up in its correct orbit because it had its own set of small thruster rockets that could be controlled from the ground. The satellite was eventually nudged into place using these rockets, but caused a delay of one year in launching the military satellite aboard the 15th flight.

Widely believed to be a spy satellite, the mystery cargo caused much controversy when the December 19, 1984 edition of the *Washington Post* reported more details about the satellite than Secretary of Defense Caspar Weinberger thought necessary. "Particularly damaging to the national security" was the way a Defense Department spokesman put it.

With an all-military crew of five, Discovery lifted off on January 24, 1985. The launch was originally scheduled for the day before, but subfreezing weather conditions forced a one-day postponement.

"We waited a day, it was a lot warmer, and we had no icing concerns whatever," said launch director Robert Sieck.

Challenger was originally slated for this flight but thermal tile problems forced the use of Discovery. In keeping with the missions's secret status, the exact time of liftoff was not announced until less than 10 minutes before it actually happened. NASA officials refused to say whether the launch had occurred on schedule.

"If that was known, that would be an indication for the Soviet Union about the mission and the type of payload we have boosted," said Air Force Lt. Col. John R. Booth. The fleet of Soviet trawlers usually found offshore for every launch was absent for this one.

The Air Force was the customer for this launch and was paying almost a third of the flight's 100-million-dollar cost. Compared to the previous launch three months before when 13,000 guests were invited, only 200 were invited for this launch, and the list was classified. Photographers were allowed to get no closer than one mile away from the launch pad.

The *New York Times* noted that it was ironic that the first secret mission had the clearest atmospheric visibility of all shuttle launches to date.

"The sky was so clear that viewers could still see the speck of light five minutes after liftoff," the paper reported.

The only bit of chatter between the ground and crew reported on this flight was just before the launch, when launch director Bob Sieck told the crew, "Have a super mission." About 2 hours after launch, NASA reported that the payload doors were swinging open. The crew was "in good spirits and has settled down to housekeeping chores," said NASA Mission Control. The agency then announced that nothing more than brief status reports every eight hours would be released.

All-Military Crew

In charge of Discovery on this flight was Navy Captain Thomas K. Mattingly, a 48-year-old veteran of the fourth shuttle flight as well as earlier Apollo flights. The rest of the five-man crew consisted of Loren J. Shriver and Ellison S. Onizuka, both of the Air Force, and James F. Buchli of the Marine Corps. All four of these men were members of the NASA astronaut corps in addition to being in the military. The fifth crewmember, Payload Specialist Gary E. Payton of the Air Force, was not in the astronaut corps. He was the first of 25 Defense Department specialists trained for duty on military shuttle missions.

When the flight ended three days later, secrecy held sway just as when the flight started. At 4:23 P.M. (EST) on January 27, 1985, Discovery landed on the concrete runway at Kennedy Space Center. In the last few moments before touchdown, NASA announced that Discovery was descending over Louisiana, the Gulf of Mexico, and making its way across central Florida.

"The crew has been welcomed home and Mission 51-C has come to a successful conclusion," said mission control after the shuttle landed.

On landing day, the restricted airspace around Cape Canaveral was again violated by four private airplanes. It was the eighth time in 15 missions that private aircraft had flown into a restricted area around the Kennedy Space Center. Since 1981, there had been 24 intrusions, 16 during launches and 8 during landings. In some cases pilots had their licenses suspended for 90 days for this infraction of flight rules.

For Commander Mattingly, it was his last spaceflight. He gave up his astronaut status to become an official of the Naval Electronic Systems Command.

Discovery lifts off from Pad 39A at 2:50 P.M. on January 24, 1985.

Discovery made a precision landing at KSC's Shuttle Landing Facility at 4:23 P.M. (EST) on January 27, 1985 to bring the 51-C mission to a successful conclusion after 3 days, 1 hour, 33 minutes, and 27 seconds of flight. Discovery touched down at approximately the 2,600-foot marker on Runway 15 and came to a stop at the 10,200-foot marker.

The 51-C astronaut crew departs Discovery after the first dedicated Department of Defense mission.

Mission Number: STS-51-D (16th Flight)   **Orbiter:** Discovery

Crew: Karol J. Bobko, Commander (*bottom left*)
Donald E. Williams, Pilot (*bottom left center*)
M. Rhea Seddon, Mission Specialist (*bottom right center*)
S. David Griggs, Mission Specialist (*top left*)
Jeffrey A. Hoffman, Mission Specialist (*bottom right*)
Charles D. Walker, Payload Specialist (*top center*)
Senator E.J. "Jake" Garn, Payload Specialist (*top right*)

Launch Prep: Orbiter Processing Facility: 54 days
Vehicle Assembly Building: 5 days
Launch Pad: 16 days

Launch from KSC: April 12, 1985; 8:59 A.M. (EST)
Due to weather problems, liftoff occurred with only 55 seconds remaining in the launch window.

Mission Duration: 6 days, 23 hours, 55 minutes, 20 seconds   **Orbits:** 110

Distance Traveled: 2 million miles

Landing: April 19, 1985; 8:55 A.M. (EST)
Kennedy Space Center, Florida (Runway 33)
One right main-gear tire had a blowout.

Wheels-Down to Stop: 10,500 feet

Mission: Canadian Anik C-1 communications satellite was successfully deployed. Leasat 3 deployment from Discovery was successful but sequencer failed to initiate antenna deployment, spin-up, and ignition of perigee kick motor. Mission was extended two days to permit crew to make certain that sequencer start lever was in proper position. Griggs and Hoffman performed spacewalk to attach "flyswatter" devices to the mechanical arm.

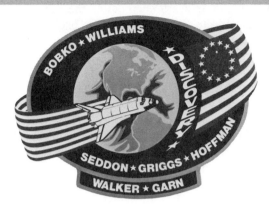

A Flyswatter and a Senator

FLIGHT 51-D, LIKE OTHERS BEFORE AND OTHERS TO COME, HAD ITS GOOD POINTS, WHERE everything went according to plan, and its bad, where even the swiftest of skilled improvisations failed to win the day. The shuttle cargo bay was filled with two satellites—Telesat of Canada's Anik C-1 and Leasat 3, a Hughes satellite leased to the Navy. The main objective of the crew was to launch these two satellites, and launch they did—but only one satellite propelled itself to its proper orbit. The other, Leasat 3, failed to go any farther after being pushed out of the cargo bay, thus leaving it in a useless orbit. What was done to try and make the helpless satellite become operational comprised the unplanned highlights of the 16th shuttle flight.

When Discovery lifted off on April 12, 1985, there were only 55 seconds left in the "window," the acceptable period for launching (determined by the position of the Earth relative to the desired orbit). The crew of seven, commanded by Karol J. Bobko, had to wait out a rainstorm but finally got off at 8:59 A.M. (EST). Others aboard were Pilot Donald E. Williams; three mission specialists, Rhea Seddon, David Griggs, and Jeffrey A. Hoffman; and two payload specialists, Charles D. Walker and Senator E.J. "Jake" Garn.

The Flyswatter

When the 15,200-pound Leasat 3 failed to propel itself to its geoosynchronous orbit 22,300 miles above the Earth (it was sent from the shuttle at a mere 200 miles up), NASA decided that if the crew could somehow close an open switch lever on the side of the satellite, its internal rockets would propel the satellite into the proper orbit. Following consultation with Mission Control, the crew was given two extra days to complete the rescue attempt. The idea was to attach some sort of snare device to the end of the mechanical arm, place the device next to the slowly spinning satellite, and trip the ignition lever that would boost the satellite to the proper orbit.

"Girl Scout camp" is the way Dr. M. Rhea Seddon described the activities on board the shuttle as the crew used their ingenuity and available materials to come up with a fly swatter-like device to trip the ignition lever. "We were measuring with tape measures and cutting and pasting and wondering what in the world this thing was going to look like when we finished up," said Seddon. Used in the construction of the flyswatter were plastic book covers, a Swiss army knife, duct tape, and scissors.

Once fabricated, the flyswatter was attached to the tip of the mechanical arm in a spacewalk by Dr. Jeffrey A. Hoffman and S. David Griggs. Dr. Seddon, who was operating the mechanical arm from inside the cabin, did flip the lever with the improvised device but nothing happened. The satellite was abandoned then and there with hopes for a future rescue attempt when astronauts would expose the inside of the satellite and do some on-site rewiring.

Slinky and Queasy

Launching satellites was not the only objective of the mission. There were also a variety of experiments to be performed. The crew played with toys such as yo-yos, jacks, and Slinkys to demonstrate (on film) to children how different space is by using objects familiar to them. The Slinky, reported Dr. Seddon, "won't slink at all. It sort of droops."

Gathering more information about motion sickness was also on the agenda for the mission. As a designated human guinea pig to test the effects of weightlessness on the human body, Jake Garn, a 52-year-old Republican senator from Utah, did not waste any time and spent the first two days of the trip suffering from nausea. Senator Garn went along as the first Congressional observer on the shuttle flight. He was acting in his capacity as chairman of a Senate subcommittee that oversaw NASA's budget.

When it came time to land, the weather was a factor, just as it had been during the launch. Discovery had to go around the globe one more time to allow a cloud cover at the Cape to move on by. Upon landing, Discovery suffered more damage than any orbiter on any previous flight. In a puff of smoke, one of the main landing gear tires blew out, and another tire was badly frayed. The brakes locked up and were damaged. After the vehicle had rolled to a stop, a burned-out basketball-sized gash was evident on a control flap. This had been caused by the loss of a heat shield tile and the resulting high temperatures that followed. More than the usual number of "impact hits" to the heat shield tiles were counted, totaling 123.

As the 16th flight came to an end, another milestone was reached in shuttle flight history. For the first time there were three shuttle vehicles on the ground at Kennedy Space Center. In addition to Discovery which had just landed in Florida, Challenger was on the launch pad ready for a flight less than two weeks away, and Atlantis had just arrived the week before and was being given the usual inspections for an upcoming flight.

Jake Garn summed up the 16th flight by speaking of what was done during the attempted satellite rescue. "Man is needed in space. You can't do the things we have done up here with unmanned space probes. You need brains, you need minds up here that can think, that are innovative."

During a simulation training flight, Sen. Jake Garn (R.-Utah) gets an initial "feel" of weightlessness as his feet float freely while he anchors himself with his hands.

Discovery and the mobile launcher are hard-down on Pad A following rollout. The 51-D mission was launched on April 12, 1985 at 8:59 A.M. (EST).

Anik C-1 is deployed from the cargo bay on 51-D's first day in space. Near the frame's center is the antenna for Leasat 3 folded against the U.S. Navy's communications satellite in a stowage position.

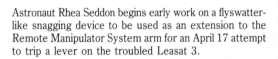

Astronaut Rhea Seddon begins early work on a flyswatter-like snagging device to be used as an extension to the Remote Manipulator System arm for an April 17 attempt to trip a lever on the troubled Leasat 3.

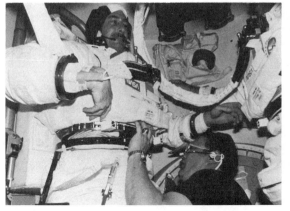

Karol J. Bobko, mission commander, assists astronauts Hoffman (left) and Griggs (partially visible at right edge) with their extravehicular mobility units prior to leaving Discovery's airlock for their extravehicular activity to deploy snagging devices on the end of the RMS.

Griggs and Hoffman join efforts to fasten one of the two snagging devices on the RMS.

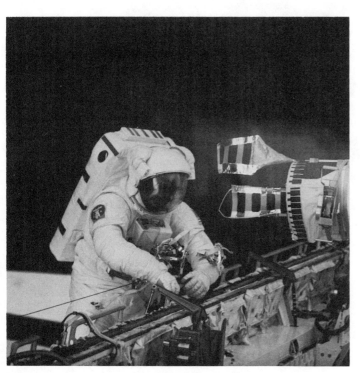

Hoffman poses next to the flyswatters.

A Flyswatter and a Senator

This aquarium-like scene came during one of the lighter moments of emergency extravehicular activity aboard Discovery. Astronaut Griggs waves from the cargo bay into the flight deck. The Earth's horizon appears both in the background of the scene and in the reflection in Griggs' helmet visor.

Astronaut Gardner, getting his turn in the Manned Maneuvering Unit, prepares to dock with the spinning Westar VI satellite. Gardner used a large tool called the apogee kick-motor capture device to enter the nozzle of a spent Westar engine and stabilize the communications spacecraft sufficiently to capture it for return to Earth.

Astronaut Williams works out on the treadmill.

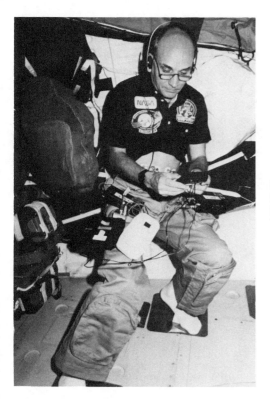

Senator Garn was scheduled as a subject for extensive medical tests. This particular experiment deals with gastric motility. Principal investigator for this experiment was astronaut Thornton, a physician.

Extravehicular activity, satellite rendezvous, and tool making weren't the only things that bore a busy appearance during the seven-day flight.

Dr. Rhea Seddon "sits" down to a meal with the help of foot restraints. Two sleep restraints are in the background.

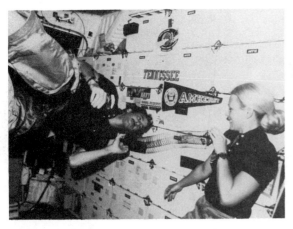

Astronauts Hoffman and Seddon demonstrate the effects of weightlessness on a Slinky.

Two of the tires on Discovery were damaged during the landing of the 51-D mission. The right outboard tire on the main landing gear was blown, and the right inboard tire was shredded. The tire damage resulted from locked brakes on the right landing gear.

Discovery landed on KSC's Shuttle Landing Facility at 8:55 A.M. (EST) on April 19, 1985. The 51-D mission lasted 6 days, 23 hours, 55 minutes, and 20 seconds.

A Flyswatter and a Senator

Mission Number: STS-51-B (17th Flight) **Orbiter:** Challenger

Crew: Robert F. Overmyer, Commander (*bottom left*)
Frederick D. Gregory, Pilot (*bottom right*)
Don L. Lind, Mission Specialist (*top left*)
Norman E. Thagard, Mission Specialist (*top center*)
William E. Thornton, Mission Specialist (*top right center*)
Lodewijk van den Berg, Payload Specialist (*top right*)
Taylor G. Wang, Payload Specialist (*top left center*)

Launch Prep: Orbiter Processing Facility: 33 days
Vehicle Assembly Building: 5 days
Launch Pad: 15 days

Launch from KSC: April 29, 1985; 12:02 P.M. (EDT)

Mission Duration: 7 days, 8 minutes   **Orbits:** 111

Distance Traveled: 2.89 million miles

Landing: May 6, 1985; 9:11 P.M. (PDT)
Edwards Air Force Base, California

Wheels-Down to Stop: 8,317 feet

Mission: First operational flight for European Space Agency-developed laboratory. Of 15 experiments scheduled aboard Spacelab 3, 14 were successful. First time animals were flown with flight crew. Two monkeys and 24 rodents were observed during mission for effects of weightlessness. Mission's main objective was to provide high quality microgravity environment for delicate materials-processing and fluid experiments. Traveled in 219-mile-high orbit inclined 57 degrees to the Equator.

7 Humans, 2 Monkeys, 24 Rats

SPACELAB 3 WAS A EUROPEAN-BUILT MODULE WHERE SCIENTISTS COULD WORK IN A SHIRTsleeve environment. The main objective of this flight was to provide a high quality microgravity environment for delicate materials-processing and fluid experiments. This was the first time that animals—2 monkeys and 24 rats—were flown aboard a space shuttle. Of the 15 experiments aboard Spacelab 3, 14 registered success. The billion-dollar module was 23 feet long by 13 feet in diameter and fitted perfectly into the cargo bay of the shuttle. The Spacelab 3 allowed direct exposure to space for experiments which required it.

This flight was called the "microgravity mission" because it was uniquely designed to produce a smooth, stable ride through space, reducing gravity and other gravity-like forces to a minimum. For the mission, delicate crystal-growth and fluid-mechanics experiments were clustered near the spacecraft's center of gravity, the most stable area of the vehicle.

For the second time in American spaceflight history, shuttle crewmembers performed scientific investigations around the clock. Two of the scientists who developed Spacelab 3 experiments conducted on-board research during the mission. Payload Specialists Dr. Lodewijk van den Berg, a materials scientist from EG&G Energy Management Corporation, Goleta, California, and Dr. Taylor Wang, a fluid physicist from NASA's Jet Propulsion Laboratory, Pasadena, California, were the second pair of career scientists to work aboard Spacelab.

Commander of the seven-member crew was 49-year-old Colonel Robert F. Overmyer, of the U.S. Marine Corps, a veteran NASA astronaut who served as pilot on the fifth shuttle mission. Assisting him was Pilot Frederick D. Gregory, colonel in the U.S. Air Force, on his first space mission.

Experiments

On 51-B, as this flight was numbered, Spacelab 3 carried 15 experiments that spanned five research disciplines: materials science, life sciences, fluid mechanics, atmospheric physics, and astronomy. Twelve of the investigations were developed by U.S. scientists, two by French scientists, and one by scientists from India.

Spacelab 3 experiments were selected by a peer review process on the basis of their intrinsic scientific merit and suitability for flight on the shuttle. NASA asked for ideas from the world scientific community. For the proposals that were accepted, a NASA working group provided scientific requirements for the actual shuttle mission. In addition, this group helped train the four Spacelab 3 payload specialists and recommended two specific individuals to perform their experiments in space.

The mission's three materials science processing experiments were to use novel techniques for growing crystals in space. Scientists believe that crystals grown in space can fare better in infrared detectors and other high technology devices.

Six of the mission's 15 experiments were to examine animal and human biological processes in the space environment, a new discipline called life sciences. Of these half-dozen experiments, four had to do with animals, namely, two dozen rats and two squirrel monkeys. The experiments were to test the suitability of their cages for more extensive animal research on later flights. In addition, a biotelemetry system was to monitor the output of sensors surgically implanted in four rats before the flight. Data on basic physiological functions, such as heart rate, muscle activity, and body temperature, were to be sent via a dedicated computer to scientists on the ground who monitored the animals' well-being. Two rats were to be killed, then examined for the effects of weightlessness, but the monkeys were to be unharmed.

On the human side, an attempt was to be made to control motion sickness through "autogenic feedback training," a technique used to voluntarily control bodily processes. The other human life-science experiment was a urine collection system attached to the shuttle's toilet for post-flight analysis of these samples. Before spaceflight, scientists could only predict how the fluids would behave in a low-gravity environment. Spacelab 3 was to give them a chance to test their theories. One of the experiments was an effort to see how microgravity could ultimately improve the processing of fluids. The idea here is that containers can contaminate fluids, and microgravity allows processing without a container. The other experiment was a model to simulate fluid flows in oceans and in planetary and solar atmospheres (gravity distorts fluid flows in experiments on Earth).

The Earth's atmosphere acts as a filter that obscures the view and study of celestial objects. Above the atmosphere, Spacelab gave instruments not only a clear view beyond the Earth, but also a better look back into the atmosphere on a global scale. The four experiments in this area were designed to take advantage of Spacelab 3's clear view. They were (1) to study the makeup of particles found in the upper atmosphere; (2) to photograph, videotape, and film the auroras or energetic particle processes occurring in the atmosphere; (3) to use a special detector to determine the composition and intensity of energetic ions from the Sun and other galactic sources; and (4) to make high quality ultraviolet images of celestial objects.

Spacelab 3 was also scheduled to launch two small satellites known as "cheapsats" (because they were built for the relatively small amount of $1-1½ million).

NUSAT, an acronym for Northern Utah Satellite, was an air traffic control radar system calibrator. It was designed to measure antenna patterns for ground-based radars operated in the United States and in member countries of the International Civil Aviation Organization. The 115-pound satellite was to have a life expectancy of about six months. NUSAT was built by Morton Thiokol Inc., Brigham

City, Utah, for a university team headed by Weber State College, Ogden, Utah, in coordination with the Federal Aviation Administration.

The other satellite, GLOMR (Global Low Orbiting Message Relay Satellite), was a data-relay communications spacecraft and was expected to remain in orbit for approximately one year. The purpose of the 150-pound, 62-side polyhedron satellite was to help the U.S. Navy locate drifting weather buoys in any ocean worldwide.

Launch Prep

Preparations for the Spacelab 3 launch began on December 13, 1983 with the arrival at Kennedy Space Center of the module used during the Spacelab 1 mission. Spacelab 1 racks and experiment equipment were removed at the KSC Operations and Checkout Building. The Spacelab module required modifications before it could be used as Spacelab 3. One was the removal of a high-quality window adapter assembly used during Spacelab 1 but not needed for Spacelab 3 investigations.

As scientific instruments and equipment arrived at KSC, they were tested and then integrated into the four single racks and eight double racks used inside the module. An atmospheric science instrument and an astronomy instrument were mounted on an experiment support structure, a lightweight carrier used to expose instruments directly to space. The support structure was moved into place behind the shell of the module, and experiment racks were moved into the module in May 1984. Because the module performed flawlessly during the first Spacelab mission, there was no need for placing it in a test stand to verify that it was compatible with the shuttle. On March 27, 1985, Spacelab 3 was transferred from the Operations and Checkout Building to the Orbiter Processing Facility and installed in the payload bay of Challenger. Spacelab experiments were operated by remote control from the Johnson Space Center's Payload Operations Control Center during a test on March 30-31, 1985.

To carry out the busy schedule of 15 scientific experiments and 2 satellite launches, the shuttle crew was divided into teams: gold and silver. Working 12-hour shifts, each team would keep the scientific investigations going around-the-clock during the seven-day mission.

The gold team consisted of Commander Robert F. Overmyer, Mission Specialist Don L. Lind, Mission Specialist William E. Thornton, and Payload Specialist Taylor G. Wang.

The silver team consisted of Pilot Frederick D. Gregory, Mission Specialist Norman E. Thagard, and Payload Specialist Lodewijk van den Berg.

Only 10 days after the last shuttle landing and 17 days after the last launch, Challenger lifted off from Kennedy Space Center at two minutes past noon on April 29, 1985, with a crew of 7 humans, 2 squirrel monkeys, and 24 rats. The launch was 2 minutes and 18 seconds late because of a malfunction of the system that controls the flow of liquid oxygen to the spacecraft. KSC officials immediately activated a secondary system, and the countdown continued again.

"It was a great ride, and we highly recommend it," reported Commander Overmyer as Challenger settled into orbit. "I don't think you'd have any trouble getting us to do it again."

This launch set a record for the shortest time between liftoffs, the old record being 34 days. Challenger also set a record for the most flights of an orbiter, seven.

Astronaut Thornton told Mission Control after he had moved into the Spacelab, "The rodents are in good shape, and the monkeys appear in good shape. One of them even came and greeted me."

The monkeys were known only as "3165" and "384-80" to avoid humanizing them. Monkeys have not always fared well as space travelers. The last one to fly on an American mission, a male macaque monkey, "Bonnie," died of hyperthermia soon after returning from an unmanned orbital flight in 1969. But one of the first two primates in space, a female squirrel monkey named "Miss Baker," not only survived her 1959 suborbital flight, but lived to the ripe old age of 27 years. Other primates in American space programs were "Able," in 1959, and "Ham" and "Enos," both of whom flew in 1961 as part of the testing preceding the manned Mercury program.

The launch day for the 17th shuttle flight was

plagued with mishaps right from the start. Although the NUSAT satellite was launched without incident 4 hours and 15 minutes after liftoff, the second satellite, GLOMR never came out of its container.

"The first deploy went fine, but we tried everything on the second, and it's just sitting there inside the can with its antennas sticking out," astronaut Thagard reported.

"No joy on the GLOMR deploy," Commander Overmyer added.

Mission Control then retorted, "You did all the right things," attempting to give the chief pilot some reassurance.

It was not immediately known what caused the failure, but there was some concern before the launch that the batteries—the one-dollar 9-volt drugstore variety used in transistor radios—might not work after prolonged time in space. By remote control, the crew was able to close the canister that contained the unlaunched GLOMR, and avoided the need for a spacewalk to tie down the lid for reentry.

Also, at their first meal the crew discovered that the water faucet would not work and had to bypass it with an all-purpose hose hookup.

Yellow Rain and Other Surprises

When work began on the 15 experiments, the trouble kept coming. The urine collection system of the life sciences experiment did not collect urine but sprayed it all over the cabin when astronauts tried to flush the toilet. After cleaning up the floating yellow nuisances, astronaut Thornton announced that he was discontinuing use of the urine collection system. Fortunately, the toilet worked when the experiment was not active.

Still another problem occurred when the astronauts changed food trays in the animal cages. Rat food particles and odoriferous monkey droppings began to float around the cabin. The crew had to again clean up a mess, this time with vacuum cleaners and plastic bags.

Despite the early troubles, the astronauts labored for the next six days to complete the experiments. At the beginning, 5 of the 15 experiments failed or refused to start, and some others were not 100-percent operable. In the end, however, the astronauts used their best fix-it skills and received ground crew assistance to chalk up 12 out of the 15 experiments as completely successful and 2 partially successful. One of the two was an experiment to study the dynamics of fluid droplets in weightlessness. The other was an instrument that counted and measured cosmic rays striking Challenger in orbit. The one experiment that completely failed was supposed to study the ultraviolet light of hot stars. A balky hatch cover prevented operation of the experiment's wide-field camera.

In all, the experiments provided a wealth of scientific data. Numerous photographs were obtained as well as 250 billion bits of computer data—enough to fill about 50,000 books of 200 pages each.

A highlight of the mission was the video taken of the mysterious polar lights called aurora borealis. The eighteen separate auroras seen and photographed by the Challenger crew were described as "spectacular and all different."

Another noteworthy experiment was Atmos, which identified and measured more than 40 different chemicals in the upper atmosphere. Scientists said that it offered enormous potential for fighting air pollution.

To Boom and Fly in L.A.

On Tuesday, May 6, 1985, after 111 orbits, Challenger returned to Earth and landed at Edwards Air Force Base in California, shortly after 9:00 A.M., local time. This was the first time that a shuttle had passed directly over Los Angeles. A double sonic boom set off burglar alarms, rattled windows, and triggered thousands of calls to the police. But, the crew, monkeys, and rats were all in fine shape.

Challenger was to have landed in Florida but concerns about possible landing gear damage on the hard concrete runway at the Cape meant that a softer lake-bed landing was in order. Also, another factor was that the Spacelab module added four tons to the landing weight, and the longer and wider landing place would be safer.

Although Challenger's tires held up in the landing, the brakes suffered some damage similar to virtually every previous shuttle flight. The brakes were removed and returned to the manufacturer for analysis.

Fifty animal rights activists, protesting the use of animals in space research, were among the spectators at the landing site on the east end of the dry lake. "We brought 'em back alive," Dr. Thornton said, apparently to answer the charges of mistreatment of animals.

"Challenger, welcome home," called out Mission Control as Commander Overmyer brought the shuttle to a stop. "Nice job, Bob."

After landing Overmyer said, "I really love Florida, but I was really happy to land Challenger here on the lake bed. I believe every rookie commander should get a shot at landing on the lake bed."

Colonel Gregory added, "I think that science is the stuff that pays for itself on these missions. It's going to improve the quality of life down here."

On landing, the Challenger had 95 small impact hits on its heat shield tiles, slightly more than average.

In answer to a request they had made from orbit the day before returning to Earth, the crew found ice cream topped with chocolate and cherries waiting to be served in their crew quarters.

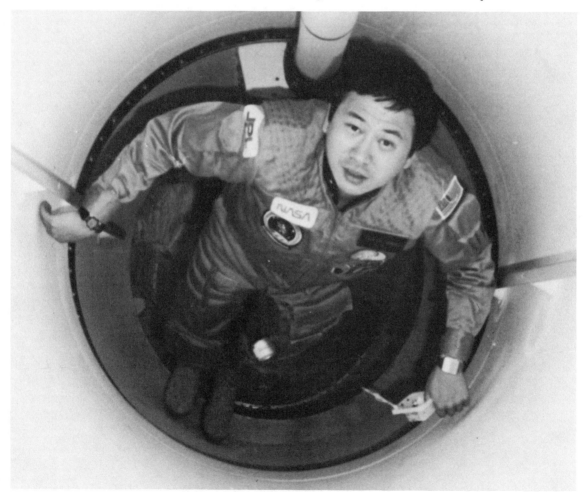

Payload Specialist Taylor G. Wang floats through the tunnelway between the mid-deck of Challenger and the long science module in its cargo bay, where the two shifts of scientists conducted research and experimentation. Dr. Wang carries a drink container in his left hand.

The 51-B Spacelab mission began with the liftoff of Challenger from Pad A at 12:02 P.M. April 29, 1985. The object of the mission was to conduct applications, science, and technology-oriented experimentation requiring the low gravity of Earth orbit and extended-duration stable vehicle attitude. The mission was planned for seven days with a landing at Dryden Flight Research Center, California, targeted for May 6 at 8:03 A.M. (PDT).

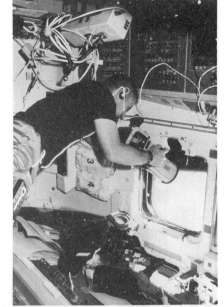

Astronaut Overmyer, 51-B mission commander, aims a Linhof camera through the flight deck windows of Challenger.

Astronaut Lind termed this scene of an aurora in the Southern Hemisphere as "spectacular," during a TV downlink featuring discussion of the auroral observation. Dr. Lind, monitoring activity in the magnetosphere at various points throughout the flight, pinpointed the spacecraft location as being over a point halfway between Australia and Antarctica. There are moonlit clouds on Earth. The horizontal band and the tall rays are aurora. The band parallel to the Earth's horizon is a luminescent of the atmosphere itself and is referred to as airflow.

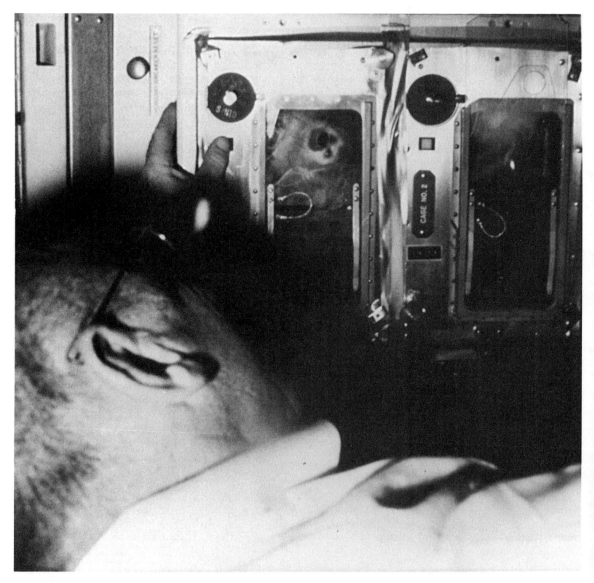

Astronaut Thornton observes one of two squirrel monkeys in the Research Animal Holding Facility on the Spacelab 3 science module.

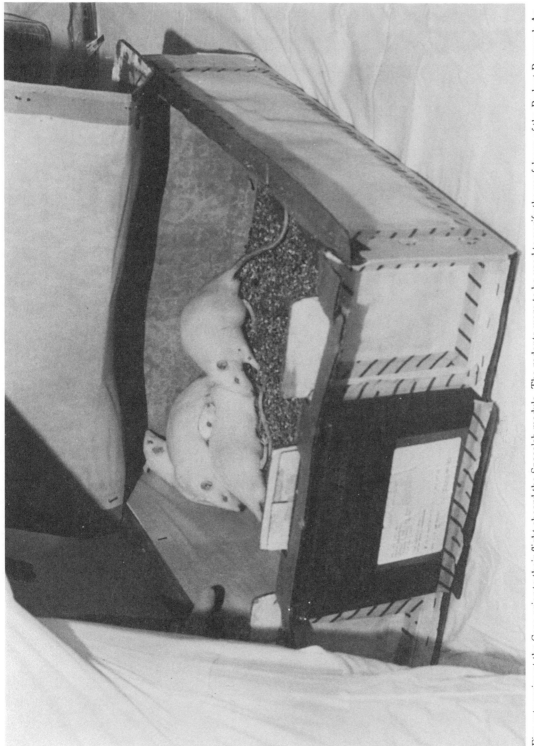

Five rats arrive at the Cape prior to their flight aboard the Spacelab module. The rodents were to be used to verify the usefulness of the Rodent Research Animal Holding Facility. Data on basic activity, feeding, and drinking levels were to be sent via dedicated computer to scientists on Earth who would monitor the animals' well-being.

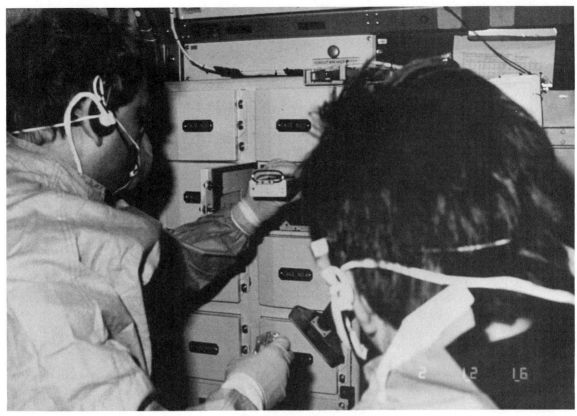
Astronaut Thagard, left, changes a tray in the Research Animal Holding Facility.

Astronaut Gregory vacuums air filters in the avionics bay. The 51-B pilot is physically located in the overhead area of the mid-deck on Challenger, but his activity is only a few meters away from the flight deck.

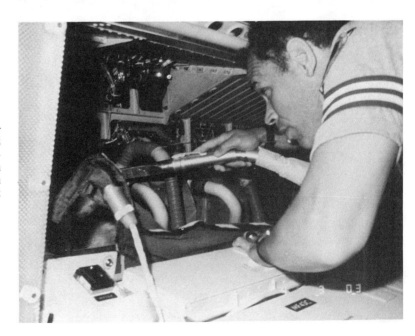

Payload Specialist Wang manipulates a 1.5-centimeter sphere in the Drop Dynamics Module (DDM) aboard Challenger. Dr. Wang was principal investigator for the experiment, developed by his team at NASA's Jet Propulsion Laboratory in Pasadena, California.

Challenger lands on Runway 17 at Edwards Air Force Base to complete a week in space for its seven-member crew and a variety of payloads. The vehicle landed May 6, 1985.

7 Humans, 2 Monkeys, 24 Rats

Mission Number:	STS-51-G (18th Flight)　　**Orbiter:** Discovery
Crew:	Daniel Brandenstein, Commander (*bottom center*) John Creighton, Pilot (*bottom left*) Shannon Lucid, Mission Specialist (*top left center*) Steven Nagel, Mission Specialist (*top left*) John Fabian, Mission Specialist (*bottom right*) Patrick Baudry, Payload Specialist (*top right center*) Sultan Salman Al-Saud, Payload Specialist (*top right*)
Launch Prep:	Orbiter Processing Facility: 38 days Vehicle Assembly Building: 7 days Launch Pad: 14 days
Launch from KSC:	June 17, 1985; 7:33 A.M. (EDT) Trouble-free countdown.
Mission Duration:	7 days, 1 hour, 38 minutes, 53 seconds　　**Orbits:** 112
Distance Traveled:	2.9 million miles
Landing:	June 24, 1985; 6:11 A.M. (PDT) Edwards Air Force Base, California
Wheels-Down to Stop:	8,130 feet
Mission:	Three communications satellites were successfully deployed: Morelos 1 (Mexico), Arabsat 1-B (Arab Satellite Communications Organization) and Telstar 3-D (AT&T). Also flown were the deployable/retrievable Spartan 1, six Getaway Special canisters, an experiment for the Strategic Defense Initiative, a materials-processing furnace, and French biomedical experiments.

An International Crew

TERMED "ONE OF THE MOST SUCCESSFUL MISSIONS OF THE SPACE PROGRAM," THE 18TH shuttle flight began with an on-time launch on June 17, 1985, and completed all that it set out to do, including a record deployment of three satellites.

With five Americans, an Arabian prince, and a French astronaut, this flight had a decidedly international cast. Even the payload was multinational, with satellites and experiments belonging to Mexico, West Germany, and the Arab world.

Commanding this flight was Daniel Brandenstein. His crew consisted of the pilot, John Creighton; three mission specialists, Shannon Lucid, Steven Nagel, and John Fabian; and two payload specialists, Patrick Baudry, and Sultan Salman Al-Saud.

Baudry, a 39-year-old French astronaut, carried out biomedical experiments similar to those flown by a French cosmonaut aboard a Soviet space mission. A native of Cameroon, Baudry was trained for scientific experiments in physiology, biology, materials processing in space, and astronomy. After joining the French Air Force Academy in 1967, he became an aeronautical engineer and logged more than 4,000 hours flying time.

Sultan Al-Saud, 28, a nephew of Saudi Arabia's King Fahd, took photographs over his homeland and participated in several experiments on this flight. The Arabian prince completed his education in the United States at the University of Denver and became a researcher in the foreign information department of the Saudi Arabian Ministry of Information. At the time of this flight he was an acting director of the Saudi Arabian Television Commercial Department. An experienced flier with a commercial pilot's license,

Al-Saud had logged more than 1,000 hours in jet aircraft and helicopters.

French Experiments

Two French experiments flew on the 18th flight. An echocardiograph experiment and a postural experiment flew as part of a cooperative project with the Centre National d'Etudes Spatiales of France. The objectives of these experiments were to obtain data regarding the response to weightlessness of the cardiovascular and sensor motor systems of the human body. The experiments were performed by French astronaut Baudry with other crewmembers participating.

On Earth the human cardiovascular system is adapted to compensate for the constant pull of gravity. During the first few days of spaceflight, astronauts' systems adapt to the sharp reduction of gravitational effects. This results in such effects as temporary pooling of blood in the head and upper torso, with changes in the size of some heart cavities, and changes in flow rates in major arteries. Data on these changes and the readaptation to gravity after the mission have important implications for crew health and safety.

The French echocardiograph used an ultrasonic technique to obtain data on these events without piercing the skin. The equipment weighed 176 pounds and was contained within two mid-deck double lockers. One double locker held the electronics; the second held the video tape recorder, the control monitor, and a stowage drawer. Payload Specialist Baudry performed this experiment twice during each flight day with help from Shannon Lucid.

The human sensorimotor functions can be categorized into four areas: muscular tone, posture, orientation, and movement. All these functional modes interact to operate within the constant field of gravity experienced on Earth. Without the physical bias and point of reference provided by gravity, these sensorimotor functions must adapt. Scientists believed that getting a better understanding of this adaptation process would help shed light on how the four functional areas interact on Earth. The French postural experiment was designed to accomplish this. Measurements of muscle activity, head movement, and up-down eye movement were taken during the flight. The measurements were obtained using biochemical electronic sensors, data tape recorders, and a camera. Baudry and the Saudi payload specialist conducted the experiment once a day during the flight. Each session took about an hour and a half.

Arabsat Communications Satellite

Arabsat was a communications spacecraft designed for launch by the shuttle and stationed at a geosynchronous orbit of 22,300 miles. The 2,800-pound satellite was boosted to its final location by a Payload Assist Module-D (PAM-D).

Morelos Communications Satellite

Morelos was one of two spacecraft scheduled for launch by the shuttle for Mexico's Secretariat of Communications and Transportation. It provided advanced telecommunications to the most remote parts of Mexico including educational TV, commercial programs over the national TV network, telephone and facsimile services, and data and business transmissions. The satellite would enable television programming to originate from at least a dozen principal cities and would allow nationwide broadcasting of cultural, educational, and athletic events.

Telstar 303

The third in the Telstar 3 series of communications satellites, Telstar 303 was capable of relaying nearly four times the number of simultaneous telephone calls commonly carried by satellites of the previous generation. This particular satellite was to replace another that was nearing the end of its service life. Improved batteries and solid-state amplifiers allowed the Telstar 3 series to operate three years longer than previous satellites—10 rather than 7 years.

The Telstar 3 series of satellites was designed by AT&T Bell Laboratories and built by Hughes Aircraft Corporation. Each Telstar 3 satellite consisted of two primary sections, containing communications units and support systems, surrounded by two concentric cylinders. Once in space, the outer cylinder dropped down about six feet, exposing the solar cells on the inner cylinder. With its antenna fully deployed in space, the satellite had an overall length of 22.4 feet and a diameter of 7.1 feet.

The two cylinders were covered with 15,588 solar cells. When exposed to the Sun, these cells—thin silicon chips—converted solar energy to electrical power to energize the satellite. When not powered by solar energy, the Telstar 3 satellite used long-life nickel-cadmium batteries.

To boost the Telstar 303 satellite into its high orbit, a PAM-D was used. Ground controllers at an AT&T facility monitored the satellite until it reached its high point, and then they fired an on-board "kick motor" which put the satellite into its final circular orbit.

Getaway Specials

Aboard the shuttle this trip, in addition to the three satellites, were six Getaway Special containers mounted along the side of Discovery's cargo bay. Three of them were of West German origin.

- Liquid Sloshing Behavior in Microgravity—This experiment examined the behavior of a liquid in a tank under microgravity conditions. It was representative of a phenomenon occurring with liquid propellants in satellite tanks. A reference fluid in a hemispherical model tank was subjected to linear acceleration inputs of known levels and frequencies. The dynamic response of the tank-liquid system was recorded and analyzed to validate and refine the mathematical models describing the dynamic characteristics of tank-fluid systems. This, in turn, supported the development of future spacecraft tanks, in particular the design of propellant management devices for surface-tension tanks. The experiment was provided by a West German firm.

- Slipcasting under Microgravity Conditions—The process of slipcasting uses a ceramic slurry to form complicated shapes of hollow bodies. On Earth, this process is limited in applications because of gravitational influences on the dispersed particles in the slurry. Sedimentation can be avoided only by use of materials with equal densities or by the use of a stabilizing additive. The latter, however, can be harmful to the desired properties of the slip-cast product.

  The goal of this experiment was to demonstrate with model materials that slipcasting was possible in microgravity, even with unstabilized suspensions using powders with different density, grain size, and concentration.

  Prior to flight, ceramic and/or metal powders were homogeneously mixed in solid paraffin. Rods of these solid slurries were then pressed into cartridges against the ends of porous ceramic rods mounted in the lower halves of the cartridges. During weightlessness, 13 samples of these solid slurries were melted. Then slipcasting was started by heating the lower parts of the cartridges. After cooling and solidification, the paraffin preserved the slip-cast layers, as well as the residual slurries, for later examination of their structure and particle distribution.

- Fundamental Studies in Manganese-Bismuth—The objective of this experiment was to produce specimens of an alloy known as manganese-bismuth, which has various magnetic applications. Materials such as manganese-bismuth could eventually lead to smaller, lighter, stronger, and longer-lasting magnets for electrical motors (in aircraft and guidance systems), surgical instruments, and transponders. The alloy was produced in a special 250-pound furnace being flown aboard the shuttle for the first time.

- Texas Student Experiments—This Getaway Special payload featured 12 different biological and physical science experiments designed by high school students from El Paso and Ysleta, Texas. The effort was supported financially by the districts involved and by businesses and other citizens in the area. The biological experiments included growing lettuce, barley, and turnip seeds, soil mold, and brine shrimp. The physical science experiments had to do with such things as the wicking of fuels and growth of crystals.

- Space Ultraviolet Radiation Experiment—This experiment was designed to measure the natural radiation field, at extreme ultraviolet wavelengths, in the upper atmosphere. The hardware consisted of a spectrometer that separated the wavelength band into two intervals of 128 discrete wavelengths. The radiation intensity at each wavelength was measured and stored on a tape recorder. The experiment instrument was developed in the Space Science Division at the Naval Research Laboratory. It was the first of a series of instruments to be developed which ultimately will have the capability of observing "ionospheric weather." It was envisioned that, someday, satellites will be stationed at high altitudes to provide global pictures of ionospheric weather conditions. Ionospheric storms, or the effects of such phenomena as solar flares or eruptions, could be monitored and their evolution followed. Effects on communication systems anywhere in the world could be observed immediately.

- Capillary Pump Loop—This experiment was the first of a thermal control system using capillary pumps. Such pumps contain no moving parts but work on the principle of capillary action. Plants and trees use this same action to transport water and nutrients from their roots to their leaves against the force of gravity. The purpose of this experiment was to demonstrate the thermal control capability of a capillary-pumped system under zero-gravity conditions, for ultimate use in large scientific instruments, advanced orbiting spacecraft, and space station components.

High Precision Tracking Experiment

The Strategic Defense Initiative Organization flew the first of a series of technology development experiments aboard the 18th flight. The High Precision Tracking Experiment was designed to test the ability of a ground laser beam director to accurately track an object in low Earth orbit. The payload for this experiment consisted of an 8-inch-diameter reflector mounted in a cylindrical housing. When removed from its storage locker, the reflector assembly was attached to the shuttle's mid-deck side hatch window in order to receive and reflect a low-energy laser beam projected from a test facility located on the island of Maui, Hawaii. The test had to be repeated because, on the first attempt, the shuttle was not facing the beam. Ground controllers had transmitted the shuttle's computer instructions in terms of feet instead of nautical miles.

Spartan 1

Spartan 1 was a spacecraft designed to be released from the shuttle, flown free in formation with the shuttle for a couple of days, and then retrieved and put back inside the shuttle for the return to Earth ("Spartan" was an acronym for Shuttle Pointed Autonomous Research Tool for Astronomy). The primary mission of Spartan 1 was to perform medium-resolution mapping of X-ray emissions from extended sources and regions, specifically, the hot gas pervading a large cluster of galaxies in the constellation Perseus. In addition, the X-ray emission from our own Milky Way was to be mapped.

Spartan 1 was a rectangular structure, 126 by 42 by 48 inches with a weight of 2,223 pounds, including 300 pounds of experiments. It was deployed and retrieved using the Canadian-built robot arm.

Landing

"Nice job, Dan. Welcome home," said Mission Control spokesman Dick Richards to Daniel Brandenstein after an early-morning landing at Edwards Air Force Base.

The only concern on landing was that, by the time Discovery had rolled more than 8,000 feet to a stop, its wheel had dug a rut six inches deep in spots on the supposedly "dry" lake bed. Officials initially thought the rut was caused by the brakes locking up, as they did in earlier flights. A later inspection showed no brake damage.

"Parts of the runway were wet and may have been a little soft," said Jesse M. Moore, the NASA associate administrator for spaceflight. There are periods in spring when the Muroc Dry Lake Bed is under water.

Referring to recent problems on other flights in getting satellites up to their correct orbits, Pilot John O. Creighton reflected on the success of the 18th shuttle flight. "Hopefully, satellite insurance rates will come down a little now."

Baudry, the French astronaut, said through an interpreter that he was particularly impressed with the firing of the twin solid rocket boosters during Discovery's launch. "It's like sitting in front of a fast train pushing you in the back. It gets through the body and into the mind for 120 seconds. Then, after the solids burn out, it becomes very smooth."

On hand to greet the Saudi prince were 10 other Saudi princes, four of them his brothers. The prince was thankful to America for the privilege of flying aboard the shuttle, but mentioned one small problem in fulfilling the dictates of his Moslem religion. "When I do my prayers, I am not able to do a complete bowing down because it is an uncomfortable form and it may cause sickness." But he had no problem facing Mecca, for it was in one direction only—down toward Earth.

Astronaut Baudry participates in a French experiment involving equilibrium and vertigo.

A low-angle 35mm tracking view of Discovery, its external tank, and two solid rocket boosters speeding from the KSC launch facility. Liftoff for 51-G occurred at 7:33 A.M. (EDT), June 17, 1985.

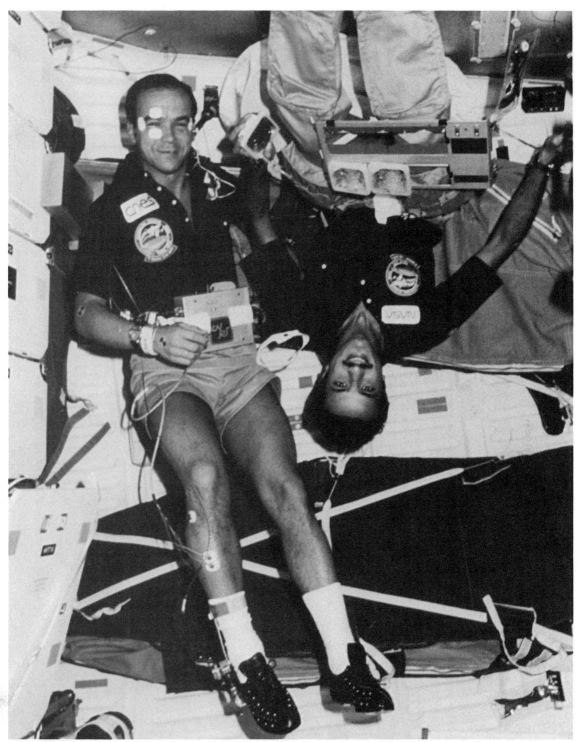
Sultan Al-Saud (right) is in the midst of a meal while Patrick Baudry conducts a phase of the French Postural Experiment.

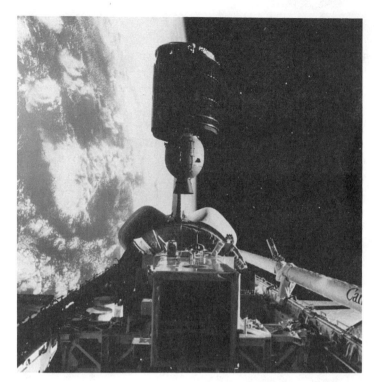

Mexico's Morelos satellite rises from Discovery to begin its life in space.

Arabsat follows suit.

The atmospheric glow of Earth, clouds, and ocean form the backdrop for Arabsat.

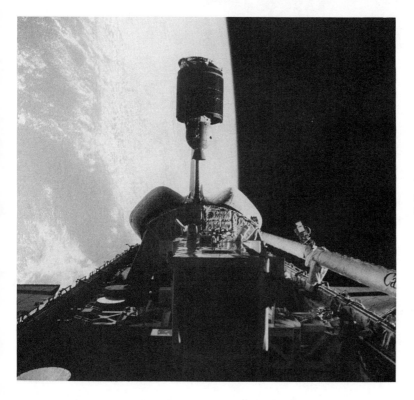

Telstar 3-D, the third and last communications satellite to be launched on 51-G, rises from the cargo bay.

International Crew

The box-shaped Spartan 1 prior to its release . . .

. . . and later as it moves out over the ocean.

Astronaut Fabian shows off a series of signs on the aft flight deck of Discovery, from whose cargo bay three communications satellites were successfully deployed. A free-flying Spartan 1 was also released and recaptured during the week-long flight.

An experiment to explore orbital photography of polarized light reflected from the Earth's surface used two Hasselblad cameras and a Dual Camera Mount to take simultaneous views with two neutral linear polarizing filters oriented in opposite planes. In this image of the Hawaiian Islands (Hawaii, at center) only the polarized light has been recorded by the film. In the highly plane-polarized sun glint upon the ocean surface, island wakes are emphasized, as currents sweep southwestward around the islands.

Payload Specialist Sultan Al-Saud assumes a posture common to the weightlessness environment of space as he logs notes.

Astronaut Creighton poses for a picture with the on-board GRID computer, which displays a likeness of Mr. Spock of Star Trek fame.

Astronaut Brandenstein shows off a Father's Day card from his daughter. The card reads, "To the world's best Dad," and is complete with artwork provided by 13-year old Adelle Brandenstein.

Discovery glides to a smooth landing on Runway 23 at Edwards Air Force Base.

International Crew

Mission Number: STS-51-F (19th Flight) **Orbiter:** Challenger

Crew: Charles G. Fullerton, Commander (*seated*)
Roy D. Bridges, Pilot (*kneeling*)
F. Story Musgrave, Mission Specialist (*center*)
Anthony W. England, Mission Specialist (*left*)
Karl G. Henize, Mission Specialist (*left center*)
Loren W. Acton, Payload Specialist (*right center*)
John-David Bartoe, Payload Specialist (*right*)

Launch Prep: Orbiter Processing Facility: 43 days
Vehicle Assembly Building: 5 days
Launch Pad: 14 days (to abort);
17 days (to launch)

Launch from KSC: July 29, 1985; 5:00 P.M. (EDT)
Emergency "Abort to Orbit" declared when Number 1 engine shut down early due to failed sensors. Launch attempt July 12, 1985 was halted at T minus 3 seconds due to malfunction of coolant valve on Number 2 engine. All three engines were shut down.

Mission Duration: 7 days, 22 hours, 45 minutes, 26 seconds   **Orbits:** 126

Distance Traveled: 3.3 million miles

Landing: August 6, 1985; 12:45 P.M. (PDT)
Edwards Air Force Base, California

Mission: Payload consisted of Spacelab 2 with "Igloo" plus 3 pallets. The primary objectives of the Spacelab 2 mission were to verify the performance of Spacelab systems, determine the interface capability of the space shuttle orbiter, and measure the environment induced by the spacecraft. Experiments covered the life sciences, plasma physics, astronomy, high-energy astrophysics, solar physics, atmospheric physics, and technology research.

False Start, Fine Ending

THE 19TH SHUTTLE FLIGHT WAS AMERICA'S 50TH MANNED SPACEFLIGHT. IT STARTED OUT as an "edge-of-your seat" ascent to orbit and ended up returning to Earth brimming with new information about the upper atmosphere, Sun, and other stars. Carried aboard this flight, 51-F, was Spacelab 2, an ambitiously complicated concoction of scientific experiments that built on the lessons learned in previous Spacelab missions during the 9th and 17th flights.

"This may be the most exciting science mission the space shuttle has flown, and we're delighted with it," said NASA shuttle chief Jesse Moore.

From the early moments of launch there were miscues, ranging from an engine that cut out too soon to a $60-million piece of equipment for pointing telescopes, which would not point. The seven crewmembers and ground control worked round-the-clock to salvage what looked at first to be a disappointing mission. When the 19th mission was over, however, over 80 percent of the mission objectives had been accomplished. Originally planned for seven days, the 19th flight was extended for one day to make up for all the time lost tinkering around with recalcitrant equipment.

Launch Preparations

Preparations for the Spacelab 2 launch began in 1982, when the European Space Agency delivered the three pallets to the Kennedy Space Center Operations and Checkout Building. These pallets were being used for the first time during the Spacelab 2 mission, but similar pallets had been tested and used successfully to perform experiments on previous shuttle missions. In 1983, the pallets were equipped with special support equipment needed for the attachment of Spacelab 2 instruments. As scientific instruments arrived at the Kennedy Space Center, they were tested and mounted on the three pallets. The "Igloo," which contained Spacelab subsystems for data collection, instrument commanding, and thermal control, was also attached to the first pallet. The cosmic ray experiment was fixed inside a special support structure located in the aft end of the payload bay.

In the fall of 1984, after undergoing many tests in Europe, the Instrument Pointing System arrived at the Kennedy Space Center. Three solar instruments and one atmospheric instrument were mounted on the pointing system, which was subsequently mounted on the first pallet.

Initial integration activities were completed in the spring of 1985 with the successful completion of testing to verify the compatibility of experiments with each other and with simulated Spacelab support subsystems. These tests culminated in the Closed Loop Test in which all commandable Spacelab experiments were operated briefly by remote control from the Payload Operations Control Center (POCC) at Johnson. The crew and scientists who developed Spacelab 2 experiments were primary participants in the integration and testing phase. Shortly after the completion of the Closed Loop Test, the Spacelab and integrated payload were placed in the Cargo Integration Test Equipment Stand to verify that they were compatible with the shuttle. This stand duplicated the mechanical and electronic systems of the orbiter.

On June 8, 1985, the Spacelab and integrated payload were transferred to the Orbiter Processing Facility and installed in the payload bay of the orbiter Challenger.

On June 12, a Spacelab-orbiter interface test was performed to check all shuttle and Spacelab 2 connections. The next day several Spacelab 2 experiments were operated again by remote control. Commands initiated at Johnson Space Center were processed through Spacelab, which was inside Challenger at Kennedy Space Center.

Launch

On July 12, 1985, all was in readiness for the 19th shuttle flight when an engine valve did not close properly and computers ordered all three of the orbiter's engines to shut down just three seconds before liftoff.

After a delay of 17 days, Challenger lifted off on July 29 with Spacelab 2 aboard. Normally, the orbiter's three main engines burn about nine minutes to get into orbit, but about six minutes into this flight, one of the engines shut down due to a failed sensor. With two engines still burning and the crew dumping rocket fuel to lighten the load, the shuttle was able to keep going but ended up about 50 miles short of the intended orbital altitude. This was the first premature engine shutdown in the history of U.S. manned spaceflight.

Had the engine failed half a minute earlier, the shuttle would not have reached orbit at all but would have had to land at one of its emergency landing sites—an airport in Zaragoza, Spain. A week later, toward the end of the mission as the shuttle was circling the globe, Charles G. Fullerton, the commander of the mission, said that he could look down and see the airport "that we almost landed at a week ago."

Once in orbit, the trouble was not over. Because the shuttle never reached its planned orbit, researchers were forced to redo many of the procedures for their experiments. During their week in orbit, the crew received thousands of additional ground commands for their experiments, a direct result of the inability to get to the right height above the Earth. A $60-million instrument pointing system did not work properly for several days and had to be reprogrammed seven times before it finally worked. One of four telescopes also failed to work, and researchers on the ground sent it hundreds of commands, but to no avail. Almost written off as a failure, the balky telescope popped on near the end of the mission.

"It's working!" shuttle astronomer John-David Bartoe exclaimed after the telescope began to record images of light.

"We're just feeling great down here that you

got it going," said George Simon in the Spacelab control center.

Spacelab 2 Scientific Investigations

This flight carried Spacelab 2, the European Space Agency's module for doing a multitude of research experiments in space. Considered one of ESA's most important programs, Spacelab represented a European commitment of almost one billion dollars. Nine European Space Agency member states—Belgium, Denmark, France, Germany, Italy, the Netherlands, Spain, Switzerland, the United Kingdom—and one state with associate member status, Australia, participated in the development of Spacelab. The objectives of the Spacelab 2 mission were to test its systems and subsystems while conducting a number of scientific investigations. In all, Spacelab 2 carried 13 experiments in seven scientific disciplines: solar physics, atmospheric physics, plasma physics, infrared astronomy, high energy astrophysics, technology research, and life sciences.

The Spacelab 2 configuration consisted of three pallets, "Igloo," and a pointing system. The three pallets were unpressurized platforms in the payload bay which, together with the pointing system, turned Spacelab into a unique orbiting observatory for studying the Sun, stars, and space environment. Igloo, a cylindrical shell attached to the first pallet, housed many of the electronics systems, such as computers and data recorders. These systems were previously located inside a pressurized laboratory module, which was not needed on this flight. The pointing system allowed for greater instrument aiming accuracy than could be attained by moving the orbiter. It was accurate enough to remain stably pointed at an object the size of a quarter from a distance of 1½ miles.

Of the thirteen scientific investigations planned for Spacelab 2, eleven of them were developed by U.S. scientists. The other two originated in the United Kingdom. The investigations were selected by a peer review process on the basis of their intrinsic scientific merit and suitability for flight on the shuttle. Proposals for experiments came through several channels, including NASA's "Announcements of Opportunity," which solicited research ideas from the worldwide scientific community. The principal investigators for each experiment then formed an Investigator Working Group, which was chaired by Dr. Eugene Urban of Marshall Space Flight Center. In addition, the working group selected and helped train the four Spacelab 2 payload specialists, and then recommended two to perform their experiments in space.

Solar Physics. Three of the mission's experiments made solar observations in visible and ultraviolet light. Above the atmosphere, the instruments saw solar emissions that were undetectable from the ground. Mounted together on the Instrument Pointing System, these instruments provided data to make a composite image of the sun's magnetic, structural, and gaseous elements. During the mission, the crew and ground investigators were able to select areas of solar activity as their viewing targets.

- Solar Magnetic and Velocity Field Measurement System/Solar Optical Universal Polarimeter (SOUP)—An instrument package of telescope and video cameras observed the Sun's magnetic field activity in different wavelengths and polarizations in visible light.

- Coronal Helium Abundance Spacelab Experiment (CHASE)—A telescope and spectrometer were used to assess solar hydrogen and helium abundance.

- Solar Ultraviolet High Resolution Telescope and Spectrograph (HRTS)—This telescope and spectrograph system observed solar radiation from the Sun's outer layers and recorded the data on film and video.

Atmospheric Physics. The atmospheric physics experiment, closely related to the Spacelab 2 solar investigations, measured solar ultraviolet radiation in the upper atmosphere. The instrument was scheduled to fly on several Spacelab missions so that long-term variations in solar ultraviolet radiation could be identified.

- Solar Ultraviolet Spectral Irradiance Monitor (SUSIM)—An instrument package of spectrometers and detectors was tuned to a narrow range of ultraviolet radiation and operated automatically every time the instrument pointing system was turned toward the Sun. Self-check calibration systems monitored the instruments and ensured accurate measurements. This instrument made a checkout flight on the third shuttle mission.

Plasma Physics. The three Spacelab 2 plasma physics experiments investigated processes in the ionosphere, the upper atmospheric region in which the shuttle travels. The ionosphere is affected by the electrified gas or plasma that streams continuously from the Sun. This mission's investigations studied the plasma environment (1) with a free-flying satellite filled with sensors, (2) by artificially stimulating the plasma with electrons, and (3) with ground observatories that can monitor the spacecraft's effect on the atmosphere.

- Ejectable Plasma Diagnostics Package (PDP)—This instrument package, also flown previously on the third shuttle mission, was extended and released by the Remote Manipulator System to make measurements after the orbiter had maneuvered to selected attitudes. On the third flight day, after about seven hours of operation as a free-flyer, the PDP was recaptured by the manipulator arm and returned to the vicinity of the payload bay. Before landing, it was locked back in place on the aft pallet. There was an understanding that if there was an anomalous situation that forced the PDP to be left behind in orbit, so be it.
- Vehicle Charging and Potential Experiment (VCAP)—The VCAP experiment was an electron generator which emitted a stream of electrons and recorded the effects of the emissions on the plasma environment. Some VCAP experiments worked with the PDP as the satellite was moved through the generated electron beam. A special television camera filmed the electron beam. This experiment also operated during the third shuttle mission.
- Plasma Depletion Experiments for Ionospheric and Radio Astronomical Studies— The effects of shuttle thruster firing on the ionosphere were measured from five radio observatories on the ground. The firings triggered chemical reactions that created ionospheric "holes." The observatories studied the changed plasma state and the transmission qualities of these altered upper atmospheric regions.

High-Energy Astrophysics. High-energy radiation in the form of X-rays, gamma-rays, and charged particles (cosmic rays) cannot be observed from Earth. Above the atmosphere, Spacelab 2 carried two large, sensitive, high-energy radiation detectors.

- Elemental Composition and Energy Spectra of Cosmic Ray Nuclei—A cosmic ray detector, on a special support structure at the end of the pallet train, was exposed to space throughout the mission. Particles entering the detector were counted and identified automatically, and the data were transmitted to the ground.
- Hard X-ray Imaging of Clusters of Galaxies and other Extended X-ray Sources/X-ray Telescope—Two telescopes, observing at different resolutions, detected distant and intense regions of X-ray emission to create X-ray images of remote clusters of galaxies and various other X-ray sources. A microprocessor system controlled target selection and pointing.

Infrared Astronomy. Infrared radiation, emitted by almost every celestial object, is best observed outside the atmosphere, where Earth's background radiation is eliminated. The Spacelab 2 telescope complemented observations made by other similar devices.

- Helium-Cooled Infrared Telescope—This small telescope measured infrared radiation from a variety of sources and could be controlled from the ground or from Spacelab computers.

Technology Research. Spacelab 2, with its delicate observational instruments, provided a chance to test advanced cooling systems. Extremely low temperatures allowed telescopes to detect celestrial radiation without the interference of background emissions from the instruments themselves.

- Properties of Superfluid Helium in Zero-Gravity—Superfluid helium is helium cooled to almost absolute zero ($-460°F$). On Spacelab 2, superfluid helium was tested for its efficiency as a cryogen (very cold refrigerant). An insulated container attached to the third pallet contained two fluid physics experiments that operated while the Shuttle was in a tail-down attitude. Sensors inside the container monitored the superfluid helium throughout the entire mission.

Life Sciences. The two Spacelab 2 life science investigations examined human and plant biological processes in the space environment. One investigation studied biochemical agents in human blood during spaceflight. The other was a variation of a plant growth experiment flown on the third shuttle mission.

- Vitamin D Metabolites and Bone Demineralization—This investigation studied the link between bone mineral loss during spaceflight and the activity of vitamin D in the human body. Blood samples were taken from crewmembers during flight, stored until landing, and then compared to samples taken before flight.
- Gravity-Influenced Lignification in Higher Plants/Plant Growth Unit—Mung beans and pine seedlings, planted in the Plant Growth Unit before flight, were monitored in flight for the production of lignin, a structural rigidity tissue found in plants. The crew checked temperatures daily and took gas samples and photographs twice during the mission.

So Many Tasks, So Little Time

Because so much time was lost in the early part of the mission trying to get all the experiments working, the crew stretched out their supplies and requested another day, primarily for more solar observation time. Mission Control happily granted the extension.

Mission Specialist Anthony England used some of the extra time to do some stargazing in a light-proof hood that kept shuttle cabin lights from blinding his view out the window.

"I'll be a little poetic here, I guess," England said. "I'm in the hood here looking at the Milky Way and the moonlit Earth going by underneath, and it is just absolutely fantastic. If anyone saw this scene they'd say it was animation, it couldn't be real."

In all, the crew of scientists collected thousands of photographs, 45 hours of video, and 230 miles of data tape.

Landing

"You guys have accomplished an enormous amount of work up there, and we've had a great deal of fun down here watching you do it," said veteran astronaut David Leestma who was working in Houston at Mission Control. "Have a safe landing."

On August 6, 1985, eight days after sputtering its way into orbit, Challenger landed at Edwards Air Force Base. About 18,000 people watched as the shuttle returned from the 19th flight.

"Everyone has collected tantalizing new data," said Dr. Eugene W. Urban, chief mission scientist. "It's going to take a long time before this data is analyzed and really fully appreciated. We've made some interesting new observations, and some have been very spectacular."

Challenger's main engines were shut down by on-board computers when redundancy was lost on a thrust chamber cooling valve on main engine #2. The shutdown occurred at T minus three seconds in the countdown for 51-F's first launch attempt on July 12, 1985.

Challenger soars upward, headed for seven days in orbit with the Spacelab 2 cargo and a seven-member crew. Launch occurred at 5:00 P.M. (EDT) on July 29, 1985.

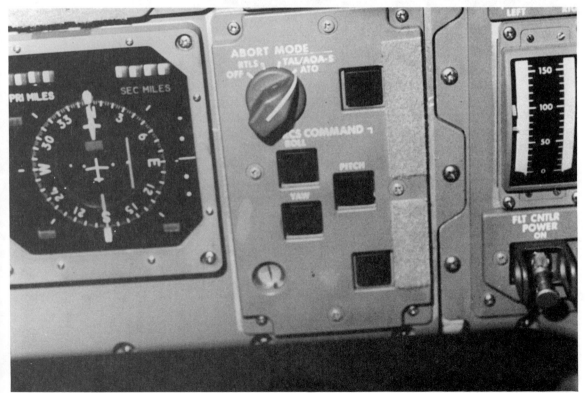

Close-up view of the display console on the forward flight deck of Challenger after its launch. The Abort Mode switch is seen in the ATO (abort to orbit) mode, marking a shuttle first.

False Start, Fine Ending

The solar optical universal polarimeter (SOUP) experiment is visible among the cluster of Spacelab 2 hardware. Various components of the instrument positioning system (IPS) are conspicuous at the center of the frame.

A close-up view of the plasma diagnostics package (PDP) held in the grasp of the end effector of the manipulator arm. Near top center is the wrist camera used for transmitting scenes from various angles.

False Start, Fine Ending

The PDP in free-flying mode.

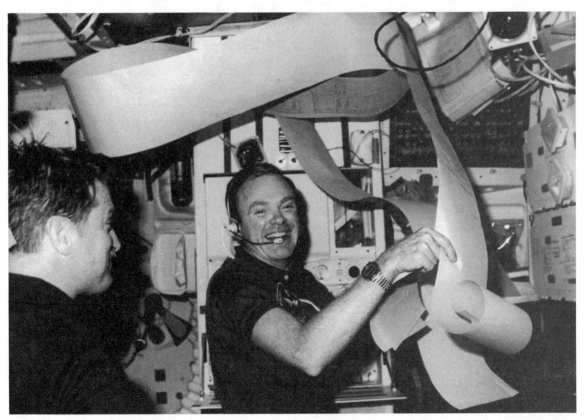
Astronauts England (left) and Bridges are surrounded by some of the prolific teleprinter copy transmitted from ground controllers. Eventually, the equivalent of several football fields' length of paper was filled with data from flight controllers.

Astronaut Henize, mission specialist, uses a 250mm lens on the 70mm Hasselblad to record a scene through the aft flight deck windows.

Loren W. Acton, 51-F payload specialist, triggers a 35mm camera recording stellar imagery through the aft flight deck overhead windows. The extension on the camera's lens is an image intensifier.

False Start, Fine Ending

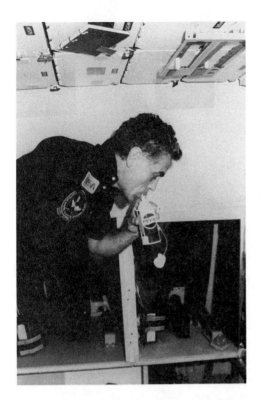

Astronaut Henize takes the Pepsi Challenge while testing one of two carbonated beverage dispensers.

Right Stuff meets Real Thing: Astronaut England tests the other beverage dispensers.

Challenger lands on the dry lake bed's Runway 23 at the end of the very successful 51-F Spacelab 2 mission.

Post-flight pictures of pine seedlings flown on 51-F. The miniature greenhouses, called Plant Growth Units, allow investigators to monitor the effect of weightlessness on the direction of plant growth and the formation of lignin, a wood substance in plants that allows them to grow upward against the pull of gravity. Plant growth in zero gravity interests NASA scientists because it may be more cost-efficient for future space colonies to grow food in space than to ship it in. However, the immediate goal was determining the effect of microgravity on production of lignin. Controlling lignin, a structural polymer that makes up 30 percent of plant tissue, is economically attractive because lignin interferes with extraction of wood fibres for the production of paper. It also reduces the use of certain plants as food because humans cannot digest it.

Mission Number: STS-51-I (20th Flight) **Orbiter:** Discovery

Crew: Joe H. Engle, Commander (*bottom left*)
Richard O. Covey, Pilot (*bottom right*)
James Van Hoften, Mission Specialist (*top left*)
John M. Lounge, Mission Specialist (*center*)
William F. Fisher, Mission Specialist (*top right*)

Launch Prep: Orbiter Processing Facility: 30 days
Vehicle Assembly Building: 7 days
Launch Pad: 22 days

Launch from KSC: August 27, 1985; 6:58 A.M. (EDT)
Launched through a hole in a storm front. Scheduled for launch August 24, scrubbed at T minus 5 minutes due to cloud system in launch area. August 25 launch attempt scrubbed at T minus 9 minutes due to failure of Discovery's Number 5 on-board computer.

Mission Duration: 7 days, 2 hours, 18 minutes, 29 seconds   **Orbits:** 112

Distance Traveled: 2.9 million miles

Landing: September 3, 1985; 6:16 A.M. (PDT)
Edwards Air Force Base, California (Runway 23)

Mission: AUSSAT-1 and ASC-1 successfully deployed on August 27, 1985. Leasat 4 deployed on August 29. Fisher and Van Hoften performed a 7 hour and 1 minute spacewalk on August 31, and a 4 hour and 26 minute spacewalk on September 1 to repair and redeploy Leasat 3 (which was first deployed from Discovery during the 16th mission in April 1985.)

Fixing Leasat 3

FOR THE SECOND TIME IN SHUTTLE HISTORY, A BUSTED SATELLITE WAS BROUGHT BACK to life and redeployed by a shuttle crew. The 20th flight, 51-I, also contained the 20th commercial communications satellite to be launched by the shuttle program in 4½ years of flights. Carrying three communications satellites, the shuttle orbiter Discovery completed another near-perfect mission. One of the reasons that the flight was not perfect was that the satellite repair activity required two spacewalks rather than one as planned.

There was some unexpected damage to the elbow controls on the Remote Manipulator Arm before launch. This meant that slower backup controls had to be used to maneuver the non-functioning satellite into position for repair. Rough weather (which later became Hurricane Elena) was speculated to have buffeted Discovery enough while on the launch pad to cause the damage. This same bad weather also caused a launch delay on the originally planned launch date of August 24, 1985. A second launch delay occurred the next day when a computer failed. Discovery finally lifted off in the early morning of August 27 during a break in that persistent storm pattern passing through Cape Canaveral. Commanded by Joe H. Engle, Discovery's crew also consisted of Pilot Richard O. Covey and three mission specialists: James Van Hoften, John M. Lounge, and William Fisher.

Repair and Salvage

Leasat 3, the silent satellite that was the repair target, had been deployed on the 16th mission four months earlier. A "sequencer" had failed, and as a result, the antenna never was put into use, and the rockets that boost the satellite never were fired. Even after extending the mission two days in an attempt to figure out how to turn the satellite on using a makeshift "flyswatter," the satellite remained silently floating in space. For this rescue mission, engineers were concerned that after floating in the cold of low earth orbit for four months, the solid rocket fuel would be cracked and, when ignited, would explode.

The general plan for putting the ailing satellite back into action was followed to the letter, except that it required two spacewalks totaling almost a dozen hours, instead of the single spacewalk originally planned. Commander Engle jockeyed Discovery to within about 35 feet of the satellite. Then, Mission Specialist Van Hoften (or "Ox" as he was known) strapped himself on to the end of the Remote Manipulator Arm, grabbed the derelict satellite with a special device, and gave the satellite to William Fisher, who was at a work station in the payload bay. Fisher performed the mechanical repair work to various parts of the spacecraft, including installation of the ground-control electronics box, a unit that deployed the antenna.

After the repair was complete, Fisher stabilized Leasat and Van Hoften installed a "spin-up" bar on the spacecraft. Discovery and the repaired Leasat 3 were maneuvered to the correct altitude for deployment. Van Hoften began to set the now-repaired satellite free by lifting the seemingly weightless spacecraft with the "spin-up" bar to start a rotation of about one revolution per minute. Commander Engle was carefully "station-keeping" Discovery while Leasat and Van Hoften kept on spinning until the satellite reached two revolutions per minute. At that point Van Hoften released the spin-up bar, and the satellite was turned over to its owners for normal operation.

Once all the instruments on the satellite were turned back on, engineers were surprised that sensors showed the solid rocket fuel to still be relatively warm, a good sign. "The temperatures we're seeing on the satellite are now higher than anybody anticipated," said Jesse Moore of NASA. "Everything about this salvage attempt has gone off like clockwork."

Commenting on this successful repair activity after the mission, Commander Engle said, "We as a crew felt very fortunate to have had the opportunity to take part in this kind of a mission that let us demonstrate how useful man can be in space. We had a great flight, and we feel it really was something that will help the space program along."

Payload

The three communications satellites deployed on this mission were AUSSAT-1, ASC-1, and Leasat 4, also known as Syncom IV-4.

AUSSAT-1 was the first of three planned Australian communications satellites. It was designed to provide domestic communications to Australia's 15-million population and to improve air traffic control services. AUSSAT, a corporation owned by the Australian government, was formed in 1982 to provide communications to Australians on a commercial basis. These communications services included direct television broadcasting to homesteads and remote communities, high quality television relays between major cities, digital data transmission for both telecommunications and business use, voice control services, and maritime radio coverage.

The deployment of AUSSAT-1 from Discovery's payload bay did not go quite as planned. A sunshield that protected the satellite from thermal damage stuck partially open. Instead of launching AUSSAT-1 on the second day of the mission as planned, NASA decided to go ahead and release AUSSAT-1 on the first day to forestall any possible thermal damage.

ASC-1 was also launched on the first day of the mission according to the original schedule. This was a shuttle first: the first time two satellites were deployed on the same day. ASC-1, owned by the American Satellite Company, provided voice, data, facsimile, and video-conferencing services to U.S. businesses and government agencies. A unique feature of this satellite was its encrypted command links, a security feature to guard against unauthorized access to the satellite command system.

Leasat 4 was released on the third day of the mission. Like earlier Leasat spacecraft, released on the 12th and 14th missions, it was leased by the

Department of Defense to replace older FleetSatCom spacecraft for worldwide UHF communications among ships, planes, and fixed facilities. Designed expressly for launch from the shuttle, this satellite was rolled out of the payload bay, rather than spun out like other satellites.

Containing their own unique upper-stage rockets, the Leasat series of satellites did not need separately purchased upper-stage rockets to boost the satellites from the shuttle deploy orbit of 200 miles to the geosynchronous orbit of 22,300 miles above Earth.

Landing

As an indication that shuttle comings and goings were becoming somewhat routine (at least in the public's mind) the smallest crowd ever to watch a shuttle landing—3,400—gathered at Edwards to see Discovery close out the 20th flight at 6:16 A.M. (PDT) on September 3, 1985. Among the spectators was Chuck Yeager, the first man to break the sound barrier (he did it in the skies above Edwards). Commander Engle was one of Yeager's former lieutenants, and Yeager said Engle was "one of the best."

Engle and his crew posed in front of the shuttle after the flight. "We as a crew felt very fortunate to have had the opportunity to take part in this kind of a mission that let us demonstrate how useful man can be in space," he said.

NASA's 20th shuttle flight may have drawn small crowds for the landing, but the $50 million it generated by launching the three satellites and repairing Leasat 3 was more revenue than the space agency had ever received for one mission.

The cargo for the 51-I mission was closed out and the ordnance work completed prior to closure of the payload bay doors. From bottom to top, the Navy/Hughes Leasat 4, the American Satellite Company ASC-1, and Australian AUSSAT-1 are securely installed in Discovery's cargo bay.

Discovery climbs skyward toward orbit on Mission 51-I. Liftoff occurred at 6:58 A.M. (EDT) August 27, 1985.

The American Satellite Company ASC-1 communications satellite rises from the cargo bay.

A single-engine orbital maneuvering system firing.

The August 28th deployment of Leasat 4.

This is one of a series of photographs covering the extravehicular activity of astronauts van Hoften and Fisher, who helped to capture, repair, and release the previously errant Leasat 3 communications satellite. Here astronaut Fisher installs a protective shield over Leasat's engine nozzle.

Here Dr. van Hoften has just given Leasat 3 a shove.

Fisher (left) and van Hoften take a break from the serious side of their spacewalk to look into Discovery's cabin.

Astronauts Lounge (left) and Covey pose from inside Discovery's cabin.

An oblique view of Hurricane Elena.

Typhoon Pat in the western Pacific. Stereoscopic overlapping photographs taken by the 51-I crew provided much more detail of the structure of cyclones than could be determined from meteorological satellite data.

Fixing Leasat 3

In this view of the Hawaiian islands of Maui (center) and Kahoolawe, smoke from the burning of sugar cane residue can be seen just west (to the left) of the cloud-shrouded Haleakala Crater.

Discovery touches down at 6:16 A.M. (PDT) on September 3, 1985.

On occasion, the ecologically unfortunate practice of oil bilging by passing ships can have scientifically useful results. The crew of Discovery recorded the boundary zone between two currents in the Mozambique Channel, revealed by the shearing of an oil bilge pattern.

Mission Number:	STS-51-J (21st Flight) **Orbiter:** Atlantis
Crew:	Karol Bobko, Commander (*bottom center*) Ronald J. Grabe, Pilot (*bottom right*) Robert Stewart, Mission Specialist (*bottom left*) David Hilmers, Mission Specialist (*top left*) William A. Pailes, Payload Specialist (*top right*)
Launch Prep:	Orbiter Processing Facility: 15 days* Vehicle Assembly Building: 17 days Launch Pad: 34 days
Launch from KSC:	October 3, 1985; 11:15 A.M. (EDT)
Mission Duration:	4 days, 1 hour, 45 minutes
Landing:	October 7, 1985; 10:00 A.M. (PDT) Edwards Air Force Base, California (Runway 23)
Wheels-Down to Stop:	8,056 feet
Mission:	Department of Defense mission. Orbital parameters and other details of the mission classified. First Atlantis mission highly successful.

*Atlantis underwent two prior processing flows in the Orbiter Processing Facility (April 14 to May 10, 1985 and May 28 to July 18, 1985) prior to beginning the final 51J processing flow on July 30. Interim periods were spent in the Vehicle Assembly Building.

ATLANTIS: Fourth Orbiter, Secret Mission

THE OUTSTANDING FEATURE OF THE 21ST SHUTTLE FLIGHT WAS NOT WHAT HAPPENED BUT the lack of information about what happened. Just as in the 15th flight nine months before, "secret," "classified," and "all-military" were words used to describe the first flight of the shuttle orbiter Atlantis, the last of the four-orbiter fleet to be built. Meager to non-existent information was the story of the four day 21st shuttle trip into space. The "highly successful" mission launched two military communications satellites.

"Of course, I can't say anything about our mission, but I can say Atlantis performed superbly its first time in space," said Karol J. Bobko, an Air Force colonel and commander of the flight. Commenting on the fact that he was the only person to have flown three of the four shuttle orbiters, he said, "I've flown Challenger, Atlantis, and Discovery now, and I think that NASA really has quite a fleet of orbiters, and we have a great national asset here."

Besides Commander Bobko, the crew included Pilot Ronald J. Grabe, Mission Specialists Robert Stewart and David Hilmers, along with Payload Specialist William A. Pailes.

On hand for the closed-to-the-public landing were fewer than a dozen reporters and 20 or so employees of NASA. Over the loudspeakers, instead of the usual cockpit-ground conversation, the "Star Spangled Banner" was played.

Strict secrecy ruled during the mission with all communications not only being blacked out but scrambled with encryption devices to keep curious Soviet eavesdroppers from tuning in. NASA broke three days of silence on the flight when it gave 24-hour notice of the planned landing at Edwards Air Force Base. Mission Control said that the crew had tested Atlantis's systems and that the orbiter had been "solid throughout the mission."

After landing and a preliminary inspection, engineers reported minor damage due to a missing insulating tile. They were trying to find out if excessive heat had reached underlying metal during the hot return through the atmosphere.

This flight was the second secret military flight in nine months, the first one being the 15th flight in January 1985.

Atlantis undergoes a Flight Readiness Firing (FRF) at Launch Pad 39A as a test of the main propulsion system. An FRF is conducted on each of the orbiters before an actual launch. All three of Atlantis' main engines had flown in space before, but had not been previously verified as a trio and with the other orbiter systems.

Atlantis, the fourth of NASA's operational orbiters, is launched on Mission 51-J. Prototype orbiter Enterprise sits adjacent to a Saturn V rocket in the foreground in a static display. 51-J was the second mission dedicated entirely to the Department of Defense.

ATLANTIS: **Fourth Orbiter, Secret Mission**

The crew of 51-J is shown departing Atlantis.

Atlantis landed at 10:00 A.M. (PDT) on October 7, 1985 on the dry lake bed at Edwards Air Force Base in southern California.

Mission Number: STS-61-A (22nd Flight)   **Orbiter:** Challenger

Crew: Henry W. Hartsfield, Commander (*front right*)
Steven R. Nagel, Pilot (*rear left*)
James F. Buchli, Mission Specialist (*front right center*)
Guion S. Bluford, Mission Specialist (*rear left center*)
Bonnie J. Dunbar, Mission Specialist (*front left center*)
Reinhard Furrer, Payload Specialist (*front left*)
Ernst Messerschmid, Payload Specialist (*rear right center*)
Wubbo Ockels, Payload Specialist (*rear right*)

Launch Prep: Orbiter Processing Facility: 61 days
Vehicle Assembly Building: 4 days
Launch Pad: 15 days

Launch from KSC: October 30, 1985; 12:00 NOON (EST)
Countdown, uneventful; ascent to orbit, normal.

Mission Duration: 7 days, 44 minutes, 51 seconds   **Orbits:** 111

Distance Traveled: 2.9 million miles

Landing: November 6, 1985; 9:45 A.M. (PST)
Edwards Air Force Base, California (Runway 17)

Wheels-Down to Stop: 8,304 feet

Mission: The first dedicated German Spacelab mission was successfully conducted from a 201-statue-mile orbit, inclined 57 degrees to the equator. Eight-member crew was largest to date. Configuration was a long module equipped with vestibular sled. Mission highlights included basic and applied microgravity research in the fields of materials science, life sciences and technology, communications, and navigation. Orbiter was controlled from Johnson Space Center with scientific operations controlled from German Space Operations Center at Oberpfaffenhofen, near Munich.

West Germany's Spacelab

As an indication of rising European influence in space, the 22nd shuttle flight dubbed "Deutschland Spacelab Mission D-1" carried a record crew of eight: five from the United States, two from West Germany, and one from the Netherlands. This was Challenger's ninth flight into space.

Billed as the first German Spacelab, the research mission was unique. For the first time in spaceflight history, the operations of a spacecraft after it arrived in orbit were controlled from somewhere other than the United States or the Soviet Union. Just outside Munich, Germany, the German Space Operations Center was the focus for the seven-day, 2.9-million-mile flight. The Germans chartered the shuttle for this flight for $64 million.

The German Spacelab D-1 was similar to prior Spacelab missions in that it was dedicated to scientific and technological research. This Spacelab came from Europe completely checked out and ready for installation. In all, there were 76 experiments conducted on a round-the-clock basis during the week-long mission. Only one experiment out of the 76 did not work. It had to do with mixing two different types of salt solution and was ruined because the mixing chamber was not properly heated.

In addition to the Spacelab experiments, another objective of this flight was to test a new system for steering the nosewheel of the orbiter. Commander Henry W. Hartsfield tested the new steering mechanism seconds after touchdown by steering 20 to 30 feet off the runway centerline and then steering back again.

"The brakes looked good coming in, and we are very, very pleased with the results of the nosewheel steering test," said Jesse W. Moore, NASA's space flight chief.

During a previous flight, brakes locked and a tire blew out on the main landing gear upon landing at Kennedy Space Center in not-uncommon crosswinds of 8-12 knots. One more successful test would allow landings back at Kennedy Space Center, eliminating the time-consuming and costly piggy-back ride aboard a 747 jet from California to Florida.

A final bit of "old business" was also taken care of on this flight. The Global Low Orbiting Message Relay Satellite was successfully launched after an unsuccessful attempt on the 18th mission in April 1985.

Background of Spacelab D-1

Spacelab D-1 was managed by the Federal German Aerospace Research Establishment for the German Federal Ministry of Research and Technology. Responsibilities included provision of the payload, payload analytical and physical integration and verification, and payload operation on orbit. Designed for experimental scientific and technological research, Spacelab D-1 was assembled over a five-year period at a cost of $175 million.

Payload Elements

The experimental facilities were arranged into "payload elements" according to scientific disciplines. Furnaces for melting, facilities for observing fluid physics phenomena, chambers to provide specific environmental conditions for living test objects, and the "vestibular sled," which exposed astronauts to defined accelerations to study the function of the inner ear, were all aboard Spacelab D-1. Most of these facilities were housed together with the necessary technical infrastructure in standard Spacelab racks within the Spacelab module.

A Sample of the Experiments

The vestibular sled consisted of a seat, for a test subject, that could be moved backward and forward with precisely adjusted accelerations along rails fixed on the floor or Spacelab's aisle. The seat was driven by an electromotor and traction rope. The sled permitted tests to investigate the functional organization of man's vestibular and orientation system and the vestibular adaptation processes under microgravity. The acceleration of the astronauts was combined with thermal stimulations of the inner ear and optokinetic stimulations of the eye.

A life sciences payload experiment package combined a group of three-element botanical or biological and two medical experiments in which a small botanical garden would be tended during the mission. Frog larvae development was investigated in one experiment. The third experiment in the field of life sciences continued the first Spacelab's medical experiments of the central venous pressure. For the first time, the internal pressure of the eye was measured. This experiment was designed to study fluid shifts under the effect of microgravity, as well as the adaptive behavior of related human organs.

Spacelab D-1 carried a navigation experiment that had two main objectives: development and testing of precise clock synchronization, and testing of a method for precise one-way distance measurement and position determination.

"We in Germany are very satisfied with this mission, and we highly appreciate the performance of the shuttle," said Herman Streuh, department head for the West German Ministry of Science and Technology.

The near-perfect record of success on Spacelab D-1 bolstered European confidence that their participation in a future U.S. space station had been assured.

Getaway Special Payload

The Global Low Orbiting Message Relay Satellite (GLOMR) was carried in a standard Getaway Special container mounted on internal ribbing of the port side of the orbiter payload bay in the vicinity of the Spacelab tunnel. Upon receiving the proper command, a full-diameter motorized door assembly on the canister opened and a spring-loaded device pushed the satellite from the container at a rate of 3½ feet per second.

The GLOMR satellite was a data-relay communications spacecraft and was expected to remain in orbit for approximately one year. The purpose of the 150-pound, 62-side polyhedron satellite was to demonstrate the ability to read signals and command oceanographic sensors, locate oceanographic and other ground sensors, and relay data from them to customers. GLOMR was carried on the 18th mission but was not deployed due to problems with the battery supply.

On December 26, 1986, the GLOMR satellite, as programmed, decayed into the Earth's atmosphere, having successfully completed its mission. Leonard Arnowitz, its project manager from Goddard Space Flight Center in Greenbelt, Maryland, reported the next day that "the GLOMR payload met all its mission objectives without a glitch." The Defense Advanced Research and Programs Agency chose GLOMR as its outstanding contract of 1986.

Landing

On November 6, 1985, Challenger made its last landing after nine successful flights. It would be launched once again two and a half months later, but it would never land again.

Challenger climbs upward toward orbit on the first dedicated German shuttle mission, 61-A.

Dutch Payload Specialist Ockels prepares to lower the eye-gear portion of the vestibular sled helmet for a test on the sled. The scientist has sensors on his face and forehead for systems monitoring.

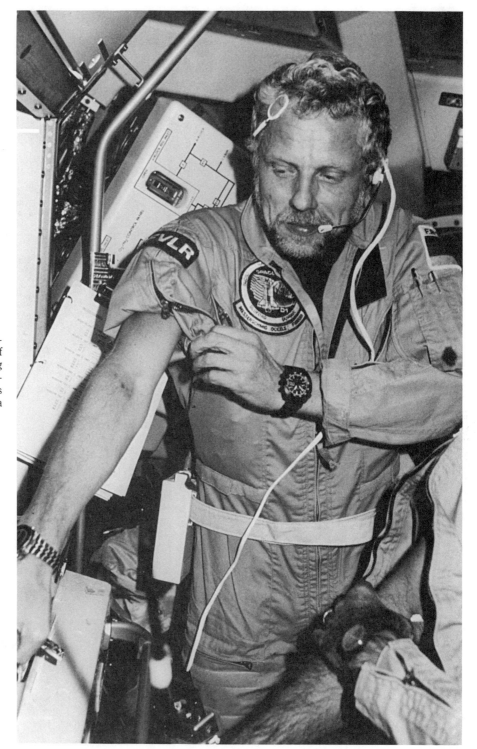

Payload Specialist Furrer shows evidence of previous blood sampling while Ockels (only partially visible) tends his right arm after having a sample taken.

West Germany's Spacelab

Payload Specialist Messerschmid, foreground, opens a door on the materials science double rack (MSDR) to begin an experiment, while Ockels performs a "run" on the vestibular sled in the background.

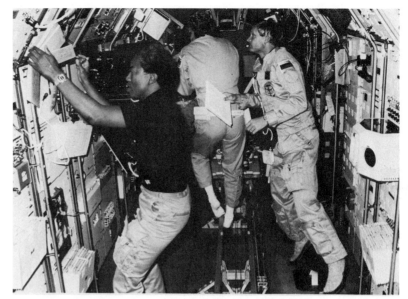

Astronaut Bluford checks work plans on a notebook near the materials science double rack while Spacelab's three payload specialists look over plans in the background. The vestibular sled (center) is captured in one of its quieter moments during the busy mission.

Eight persons returning from space at one time—a record which was set at 9:45:39 A.M. (PST) on November 6, 1985 when Challenger's wheels came to a stop on the runway at Edwards Air Force Base. The main gear are seen touching down at 9:44 A.M.

Mission Number:	STS-61-B (23rd Flight)
Orbiter:	Atlantis
Crew:	Brewster H. Shaw, Jr., Commander (*kneeling; right*) Bryan D. O'Connor, Pilot (*kneeling; left*) Mary L. Cleave, Mission Specialist (*top center*) Sherwood C. Spring, Mission Specialist (*top right center*) Jerry L. Ross, Mission Specialist (*top left center*) Rodolfo Neri, Payload Specialist (*top right*) Charles Walker, Payload Specialist (*top left*)
Launch Prep:	Orbiter Processing Facility: 27 days Vehicle Assembly Building: 4 days Launch Pad: 14 days
Launch from KSC:	November 26, 1985; 7:29 P.M. (EST) The second night shuttle launch.
Mission Duration:	6 days, 21 hours, 4 minutes, 50 seconds
Orbits:	109
Distance Traveled:	2.8 million miles
Landing:	December 3, 1985; 1:33 P.M. (PST) Edwards Air Force Base, California (Runway 22)
Mission:	Three communications satellites were successfully deployed, Morelos-B (Mexico), AUSSAT-2 (Australian) and Satcom Ku-2 (RCA). Other significant activity included two experiments to test the feasibility of assembling erectable structures in space. These experiments were EASE (Experimental Assembly of Structures in Extravehicular Activity) and ACCESS (Assembly Concept for Construction of Erectable Space Structures). These required two spacewalks by Ross and Spring. The spacewalks lasted 5 hours, 32 minutes, and 6 hours, 38 minutes. The Continuous Flow Electrophoresis System was flown for the third time and operated by Walker. Neri was the first Mexican national to be flown in space. Orbital altitude: 218 to 235 statute miles with inclination of 28.5 degrees.

Building Space Structures

LIKE MILLIONS OF AMERICANS ON EARTH BELOW THEM, THE CREW OF THE 23RD SPACE SHUTtle flight were taking part in "holiday travel." Commander Brewster H. Shaw Jr. and his crew of half a dozen others were aboard the orbiter Atlantis for the first Thanksgiving spaceflight since the third Skylab mission in 1973.

The highlight of the mission was the snapping together of a 45-foot tower composed of metal struts that were stored in the payload bay. This practice demonstration was for the space station that will be built in a similar fashion in the 1990s.

Three Satellites

After a night launch two days before Thanksgiving, Atlantis began its week-long flight by releasing three satellites. One of them, the RCA Satcom Ku-2, stood out from all previous commercial satellites. It was not insured. In addition, it was put into high orbit using a previously untried rocket motor.

RCA officials did not insure the craft because the insurance premiums were prohibitively high due to earlier satellite losses. The new rocket motor, a Payload Assist Module, was a more powerful one designed to lift up to 4,200 pounds into orbit as compared to the usual 2,800 pounds. Satcom Ku-2 was the first of three satellites to be launched for RCA to provide television and educational services to multi-unit residential complexes, such as condominiums and apartments, hotels, hospitals, and schools. In addition, it would provide direct-to-home services for a wide range of programming choices to homes far removed from cable systems and standard over-the-air television stations.

The second in a series of communications satellites for Mexico was launched on this flight. Morelos-B provided the most remote parts of Mexico advanced telecommunications: educational TV, commercial programs over the national TV network, telephone and facsimile services, and data and business transmissions.

For Australia, AUSSAT-2 was launched. It provided direct television broadcast to homesteads and remote communities, high quality television relays between major cities, and other services. AUSSAT-1 was launched on the 20th shuttle flight.

By mealtime on Thanksgiving day, the crew had successfully launched all three satellites. "That's three for three," said Mission Control. "Now we can look forward to some ground-breaking with your new construction technique tomorrow."

Turkey in Orbit

Circling the globe about 230 miles up, the crew of Atlantis celebrated the Thanksgiving holiday with a meal of smoked turkey, chicken consomme, cranberry sauce, green beans, corn, pasta, butter cookies, and lemonade. The turkey had been irradiated with gamma rays to preserve it. The consomme and vegetable mixture were freeze-dried. The butter cookies and lemonade were the real items.

Payload Specialist Rodolfo Neri, Mexico's first person in space, also carried tortillas and high protein cookies made with amaranth flour from Mexico. He was on board to complete several experiments of Mexican design.

Mission Control relayed a message from Kathleen Shaw, wife of mission commander Brewster Shaw. "Your families on Earth wish their loved ones aboard Atlantis a happy Thanksgiving," said the message.

"This is one of the best Thanksgivings all of us have spent. We hope the same is true for all of you down there, including our loved ones and friends," replied Shaw.

Construction in Space

Demonstrating different techniques for building structures in space was a major purpose of this mission. Two structures, one a tower and the other a pyramid, were built during two separate six-hour spacewalks by Mission Specialists Jerry L. Ross and Sherwood C. Spring. Known as EASE/ACCESS, the experiments showed that small components could be used to form larger structures, just as will eventually be done to build the space station. EASE stood for Experimental Assembly of Structures in Extravehicular Activity, and ACCESS was for Assembly Concept for Construction of Erectable Space Structures.

The first spacewalk was devoted to experiments to study human performance of construction tasks in space. The two astronauts were videotaped as they did their work. These tapes were later used in time-and-motion studies of their activity.

The second spacewalk was dedicated to supplementary experiments that explored alternative construction techniques. The specific objectives were to:

- gain valuable construction experience while in orbit;
- compare assembly rates and techniques used in space to those used during ground assembly test in neutral buoyancy water tank tests simulating the space environment;
- identify ways to improve erectable structures to ensure productivity, reliability, and safety; and
- evalute space station assembly and maintenance concepts and techniques.

Both EASE and ACCESS were large space structures distinguished by different assembly methods and physical characteristics.

ACCESS was a 45-foot, high-rise tower composed of many small struts and nodes. The entire structure consisted of 93 tubular aluminum struts of one-inch diameter. The tower was assembled with crewmembers in fixed work stations. EASE was a geometric structure that looked like an inverted pyramid and was composed of a few large beams and nodes. Crewmembers Spring and Ross moved about during the pyramid assemblies rather than working in fixed positions.

No tools were used for building the tower and the pyramid. The crewmembers snapped together the prefabricated components to form each structure. Both the larger pyramid beams and the smaller tower struts were joined by nodes, clusters of sockets which were locked into place by sleeves on the ends of the beams or struts.

The construction demonstration was supported by five of the seven crewmembers: the commander, the pilot, and three mission specialists. The ability to make timely, on-the-scene judgments and to provide activity reports was central to the first orbital construction mission.

A third mission specialist, Mary Cleave, operated the orbiter's robot arm to position crewmembers during some special construction tasks. To measure learning and productivity, the structures were assembled and disassembled repeatedly.

In addition to building and taking apart the structures, the spacewalk construction duo maneuvered them to gain experience handling large frameworks. Mission Specialist Sherwood Spring twirled the 12-foot pyramid around in the weightlessness of space. "I could balance this thing on my nose," he said.

To simulate the stringing of electrical cable, both crewmembers strung a flexible cable along the tower's truss framework.

Data Collection

Careful observation of crew activities was needed to understand the human factor elements of space construction. Aspects such as learning, productivity, and fatigue were important, as well as the biomedical effects of working during spacewalks. Video cameras located in the orbiter payload bay and on the robot arm were used to record images of each crewmember at work. Movie cameras mounted in the aft flight deck windows were synchronized to generate a 3-D film, which was to be used later to analyze body positions, equipment locations, and any difficulties in completing a task.

Other Activities

The Continuous Flow Electrophoresis System flew again on the 23rd flight. This mission tested the mass production concept. Approximately one liter of raw hormone material was purified during the first five days of flight. Payload Specialist Charles Walker performed sample evaluations throughout the flight, and upon return to Earth, the material was submitted to the Food and Drug Administration for testing.

Also being flown again was the Diffusive Mixing of Organic Solutions, designed to grow crystals through the combination of organic solutions. This experiment was flown on the 14th flight in November 1984. Under the supervision of astronaut Mary Cleave, the mixing apparatus built several types of organic crystals, which were larger and more sure than those grown in a positive gravity environment.

The crew on this flight took a series of photographs of Ethiopia and Somalia, which had been suffering from a period of extreme drought and famine. Experts hoped that, by looking at pictures taken from space, they could find signs of subterranean water deposits.

Getaway Special

To stimulate Canadian student interest in the space program, a contest of science experiments was held for high school students across Canada. Nearly 300 students competed, with 72 entries. The winner, entitled "Towards a Better Mirror," proposed to fabricate mirrors in space that would provide higher performance than similar mirrors made on Earth.

Cleave's Comet

A little past halfway into the week-long flight, Mission Specialist Mary Cleave created something of a sensation back on Earth during a routine dumping of wastewater from Atlantis. The wastewater was dumped at just the right time and place over Houston for the Sun to illuminate not only the orbiter Atlantis, but also the water droplets streaming out of it.

"People saw both the shuttle and the dump. The stream appeared to be about 15 miles long, and it was very bright. We're calling it Cleave's Comet," said astronaut Sally Ride, who observed the phenomenon from the ground in Houston.

Landing

Closing out its second flight, Atlantis landed on the concrete runway at Edwards Air Force Base in the early afternoon of December 3, 1985. The "dry" lake bed, where the shuttle usually lands, was damp from recent rains.

"Welcome home, Atlantis," said Mission Control. "Great landing."

The cargo for Mission 61-B is loaded into the payload canister at the Vertical Processing Facility in preparation for delivery to Pad 39A and installation into Atlantis' cargo bay. From bottom to top, AUSSAT-2 (Australia), Morelos-E (Mexico), Satcom Ku-1 (RCA), and the EASE-ACCESS experiments. EASE and ACCESS were designed to test the feasibility of assembling erectable structures in space.

Atlantis is towed to the Vehicle Assembly Building for mating with the external tank and boosters.

Atlantis ascends into the night at 7:29 P.M. (EST), November 26, 1985.

Astronaut O'Connor photographed this scene of astronaut Ross's EVA egress. Silhouetted in the foreground are part of the airlock and its hatch.

Astronaut Walker works with the protein crystal growth experiment—one of a series of tests to study the possibility of crystalizing biological materials. Walker rests the experiment against the larger Continuous Flow Electrophoresis System.

Building Space Structures 251

Payload Specialist Neri opens a stowage drawer to begin an experiment. Evidence of recent photographic work (used rolls of film) adorn the stowage locker area along with camera equipment and other paraphernalia.

Neri's alma mater crest and a Morelos insignia adorn a locker.

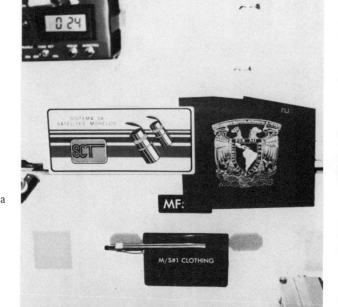

252 FLIGHT 23: ATLANTIS

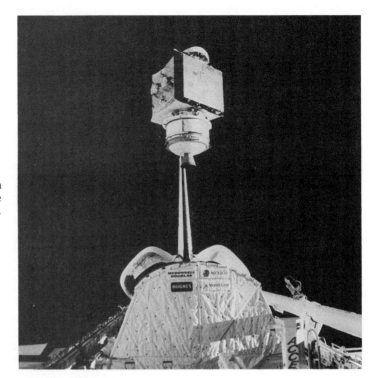

The 4,444-pound RCA Satcom Ku-2 communications satellite moves quickly away from Atlantis.

The deployment of Morelos-B, the second in a series of communications satellites for Mexico, was initiated by astronaut Spring on November 27.

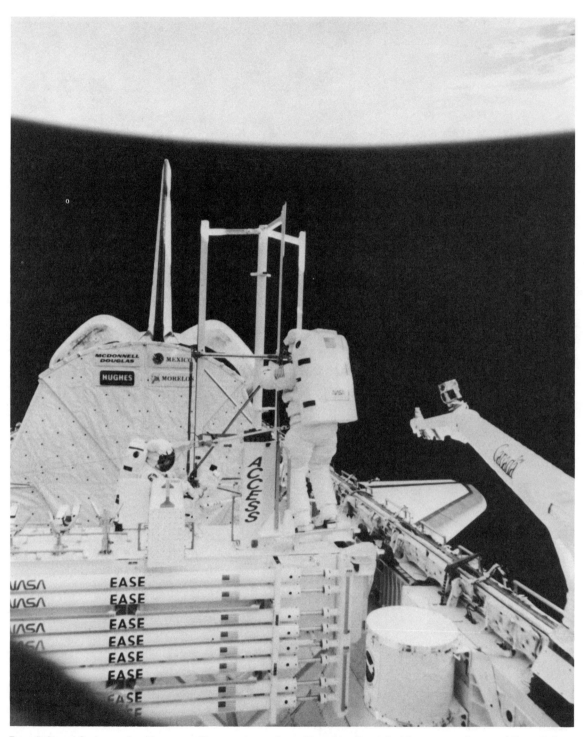
Ross (left) and Spring work with ACCESS. The remote arm is positioned to allow television cameras to record the activity.

Astronauts Ross (bottom) and Spring assemble the EASE project.

Building Space Structures 255

Astronaut Spring on the end of the manipulator arm checks joints on the Assembly Concept for Construction of Ejectable Space Structures (ACCESS) tower extending from Atlantis's cargo bay. Scattered clouds over the Gulf of Mexico form the backdrop for the scene.

Atlantis touched down on a concrete runway at Edwards at 1:33 P.M. (PST), December 3, 1985.

Building Space Structures

Mission Number: STS-61-C (24th Flight)   **Orbiter:** Columbia

Crew: Robert L. Gibson, Commander (*bottom right center*)
Charles F. Bolden, Jr., Pilot (*bottom left center*)
Franklin Chang-Diaz, Mission Specialist (*bottom right*)
Steven A. Hawley, Mission Specialist (*top center*)
George D. Nelson, Mission Specialist (*top right*)
Robert Cenker, Payload Specialist (*bottom left*)
Congressman Bill Nelson, Payload Specialist (*top left*)

Launch Prep: Orbiter Processing Facility: 108 days
Vehicle Assembly Building: 10 days
Launch Pad: 35 days

Launch from KSC: January 12, 1986; 6:55 A.M. (EST)   **Orbits:** 96
Seven launch delays over a 25-day period.

Mission Duration: 6 days, 2 hours, 4 minutes, 9 seconds

Distance Traveled: 2.2 million miles

Landing: January 18, 1986; 5:58 A.M. (PST)
Edwards Air Force Base, California
Landing attempt on January 16 waved off due to unfavorable weather. On January 17, unfavorable weather forced another wave-off. Mission extended by one day to provide a Kennedy landing opportunity on January 18 and avoid time lost in an Edwards landing and turnaround. Landed at Edwards after third weather wave-off in Florida.

Mission: RCA Satcom Ku-1 successfully deployed. Attempt to take light-intensified pictures of Halley's Comet failed due to battery problems.

Last Complete Flight

THE FIRST OF 15 PLANNED SHUTTLE FLIGHTS FOR 1986 WAS A RECORD SETTER. NEVER BEfore were there so many false starts to launch or wave-offs for landing. Seven launch delays over a 25-day period were caused by weather and technical problems.

Columbia's return to Earth was two days late, at night, and in California, after three attempts to land in Florida. The delays had NASA concerned because Columbia's next flight, set for March 6, featured a rare look-see at Halley's Comet, and timing of the launch date was critical. The March 6 date was to coincide with Soviet and European unmanned probes approaching the comet.

"We finally started a very busy year," said Bob Sieck, shuttle operations director, on January 12 as the launch control team cheered at Kennedy Space Center. "The launch team is very tired, but very happy and very proud." The previous record for delays was held by Discovery, with six delays before it got off the ground on its first flight in August 1984.

Overhaul

This was Columbia's first trip in two years, and it was not the same ship that had inaugurated the shuttle era in 1981. To bring it up to the same specification levels as later orbiters in the fleet, Columbia underwent an 18-month, $42-million overhaul at the factory where it was built. Columbia received new instruments and updated navigation equipment, and its wings and body were strengthened. The factory also took out the existing ejection seats and replaced them with lighter seats. The overhaul, however, did not include any changes to the rocket engines.

Several experimental devices were added throughout the spacecraft, one in the tail fin. A cylindrical housing known as the Shuttle Infrared Leeside Temperature Sensing Experiment (SILTS) was designed to obtain high-resolution infrared images of the leeside (upper side) surfaces of Columbia's left wing and fuselage as the orbiter reenters the Earth's atmosphere. The images were for making detailed temperature maps that would indicate the amount of aerodynamic heating of those surfaces in flight.

Another less obvious change to Columbia was a new nose cap to house the Shuttle Entry Air Data System (SEADS) experiment. A number of pressure sensors inside the nose cap provided aerodynamic flight characteristics during reentry.

An Upper Atmosphere Mass Spectrometer (SUMS) experiment was designed to sample air at Columbia's surface through a small hole. Inside the nosewheel well, SUMS was to collect molecules of various gases, allowing determination of orbiter aerodynamic characteristics at altitudes where the atmosphere is extremely thin.

On ascent the actual load on the wings was to be calculated using transducers located on the top and bottom sides of the wings. This would help find out if better performance could be gained from the orbiter.

Delays

- December 18, 1985: Ground crews fell behind schedule and were given an extra day to catch up.

- December 19, 1985: Countdown was stopped with 14 seconds to go, due to out-of-tolerance turbine reading on one of the solid rocket boosters.

- January 4, 1986: NASA decided on a delay to give astronauts, just back from the holidays, time to freshen up their training on flight simulators.

- January 6, 1986: Countdown was stopped with 31 seconds to go, due to a problem with the fuel fill-and-drain valve.

- January 7, 1986: Countdown was held while launch team waited for weather at two landing sites to clear. Weather at Kennedy Space Center was also marginal. Launch was postponed until January 9.

- January 8, 1986: Launch was rescheduled from January 9 to January 10 after an obstruction was found in a main engine valve.

- January 10, 1986: Launch attempt was abandoned due to heavy rains and driving wind in launch pad area.

Prior to four of the seven launch delays, the crew boarded the spacecraft. The other three delays were announced before the crew went aboard.

Launch

On January 12, at 6:55 A.M. (EST), Columbia finally lifted off with its crew of seven. Aboard for this flight was the congressman whose district covers Kennedy Space Center, Bill Nelson. Also aboard was Franklin Chang-Diaz, the first Hispanic-American to go into space. Robert L. "Hoot" Gibson commanded this mission, and the pilot was Charles F. Bolden, Jr. Mission Specialist Steven A. Hawley took honors on this trip as the most delayed astronaut; he was aboard the six-time delayed first flight of Discovery and boarded Columbia wearing a Groucho Marx disguise.

"What we were attempting to do was keep Columbia from being able to tell who was getting on," said Commander Gibson.

Even though much delayed, the launch on January 12 went without a hitch into a clear sky. "It was the most beautiful [launch] I have ever seen, and I've been to most of them," said Grace Nelson, wife of the politician-crewmember.

Mission Operations

Once into orbit the crew let go the single satellite carried aboard for this mission. The $50-million RCA Satcom was the second in a planned fleet of three communications satellites. The first of the series was deployed a few months earlier in November 1985.

The crew then turned their attention to the other objectives of the flight, which included more than a dozen experiments. One of them was an attempt to take pictures of Halley's Comet using a light-intensifier device strapped to a camera; it was designed to make Halley's Comet many times brighter for a better photograph. The photographs of the comet were to be the first taken from space while the comet was so near to the Sun. However, someone inadvertently left the batteries on when the device was stowed aboard, and the batteries were dead. Mission Specialist George D. Nelson replaced the batteries, but the light intensifier still did not work. He then just took regular 30-second exposures of the comet and hoped for the best.

"Without the intensifier, the comet is difficult to find," Nelson said, "but I think we got it. We got five different exposures, but I'm not sure how bright they're going to be. Maybe we got lucky. I don't know."

One experiment that worked well was a measurement device in the payload bay to detect dust particles and other debris that could interfere with the operation of infrared telescopes. The Particle Analysis Cameras for the Shuttle (PACS) experiment provided film images of particle contamination around the shuttle in support of future Department of Defense infrared telescope operations.

Other projects aboard Columbia included experiments in materials science, and medicine. Thirteen Getaway Special containers were also carried on this flight.

Trouble Coming Down

After being in orbit for four days, Columbia was told to come home to a January 16 landing in Florida. This was a day short of the original plan because of the tight timetable for launches planned for 1986. Fog and rain prevented this, and the crew was waved off to try again the next morning. The crew used the extra time to continue their experiments.

It was the same story on the morning of January 17 as Columbia was waved off again due to bad weather. "It looks like you're going to get your five days," said Mission Control.

NASA really did not want to have Columbia land at Edwards because of the six days it would take to ferry the orbiter from California to Florida—six valuable days, especially if there was to be a launch on March 6. The crew took this iffy situation in good spirits. They even broke into song. They got on the radio and sang their rendition of "Where or When" with their own lyrics:

. . . The clothes we're wearing are the clothes we've worn, the food that we are eating is getting hard to find since we can't remember where or when.
. . . Some things that happen for the first time seem to be happening again.
. . . And so it seems that we will be old at burn [braking engine firing], return to Earth and land somewhere, but who knows where or when.

Early on January 18, Columbia was waved off a third time from Kennedy. That meant that not only would it land at Edwards but the landing would take place in darkness (a night landing had only taken place once before). The Florida landing, if it had taken place, was to be a 7:31 A.M. (EST), and the landing at Edwards was 87 minutes later at 5:58 A.M. (PST).

Other than the fact that it was dark, landing conditions at Edwards were perfect. Six floodlights shining with 4.8-trillion candlepower showed the way.

"It sure took us a number of tries to get up in the air, and it sure took us a number of tries to get back down, but it was all completely worth it," said Robert Gibson, commander of the mission.

It's a bird, it's a plane, it's Congressman Bill Nelson training in the "zero-gravity" aircraft.

Columbia rises into the early morning sky over Cape Canaveral, leaving a spectacular trail of exhaust.

Last Complete Flight

The Satcom Ku-1 communications satellite spins from the protective shield in Columbia's aft cargo bay and begins to rise into space, headed toward several maneuvers which will eventually put the spacecraft in its desired position. The materials science laboratory is visible in the foreground.

Congressman Nelson uses a unique pair of goggles to participate in one of a number of detailed supplementary objective studies for NASA's Space Biomedical Research Institute. The device pictured is a pair of ocular counter-rolling goggles.

Astronaut Bolden, 61-C pilot, mans the pilot's station on Columbia's flight deck prior to reentry.

Mission Number: STS-51-L (25th Flight) **Orbiter:** Challenger

Crew: Francis R. Scobee, Commander (*fourth from left*)
Michael J. Smith, Pilot (*second from right*)
Judith A. Resnik, Mission Specialist (*third from left*)
Ellison Onizuka, Mission Specialist (*right*)
Ronald E. McNair, Mission Specialist (*third from right*)
Christa McAuliffe, Payload Specialist and Teacher-in-Space (*left*)
Gregory Jarvis, Payload Specialist (*second from left*)

Launch Prep: Orbiter Processing Facility: 35 days
Vehicle Assembly Building: 6 days
Launch Pad: 37 days

Launch from KSC: January 28, 1986; 11:38 A.M. (EST)
Several launch attempts were made between the originally scheduled date, January 22, and January 28. This was the first space shuttle launched from Complex 39-B. All prior shuttle flights were launched from Pad A, An explosion 73 seconds after liftoff claimed the crew and vehicle.

Mission Duration: 73 seconds

Mission: The objectives of this mission included the deployment of Tracking and Data Relay Satellite-B (TDRS-B) and the flying of the Spartan Halley experiment. Spartan was to have been deployed from Challenger's payload bay to bring two ultraviolet spectrographs to bear on the coma and tail of Halley's Comet. This mission also included the Teacher-in-Space project.

The Challenger Disaster

"WE HAVE A REPORT FROM THE FLIGHT DYNAMICS OFFICER THAT THE VEHICLE HAS EXploded," said Stephen A. Nesbitt, the voice of Mission Control in Houston.

With those words came the first official recognition of what thousands had just seen in person and millions had just seen on television. Challenger's last short flight of 73 seconds had ended in disaster, on January 28, 1986, killing all seven crewmembers. The names of the dead astronauts, Francis R. Scobee, Michael J. Smith, Judith A. Resnik, Ellison Onizuka, Ronald E. McNair, Christa McAuliffe, and Gregory Jarvis quickly became known across America as the nation mourned its worst accident in spaceflight history.

Challenger's fateful last flight, 51-L, was originally scheduled for January 22 but was delayed a number of times. At 11:38 A.M. when Challenger left Launch Pad 39 B at Kennedy Space Center, the temperature at ground level around Challenger was 36°F. This was 15° colder than that of any previous launch. On-board in the payload bay for this flight was another tracking satellite and an experiment to explore Halley's Comet. Christa McAuliffe was flying as America's first Teacher-in-Space, part of a program to introduce ordinary citizens to space travel. She was to conduct two lessons from space for millions of students nationwide.

The Commission's Report

The Rogers Commission, which studied the accident for several months, provides the following, somewhat technical, description of what happened after launch:

At approximately 37 seconds, Challenger encountered the first of several high-altitude wind shear conditions, which lasted until about 64 seconds. The wind shear created forces on the vehicle with relatively large fluctuations. These were immediately sensed and countered by the guidance, navigation and control system.

Although flight 51-L loads exceeded prior experience in both yaw and pitch planes at certain instants, the maxima had been encountered on previous flights and were within design limits.

The steering system of the solid rocket booster responded to all commands and wind shear effects. The wind shear caused the steering system to be more active than on any previous flight.

At 45 seconds into the flight, three bright flashes appeared downstream of Challenger's right wing. Each flash lasted less than one-thirtieth of a second. Similar flashes have been seen on other flights. Another appearance of a separate bright spot was diagnosed by film analysis to be a reflection of main engine exhaust on the Orbital Maneuvering System pods located at the upper rear section of the orbiter. The flashes were unrelated to the later appearance of the flame plume from the right solid rocket booster.

Both the shuttle main engines and the solid rockets operated at reduced thrust approaching and passing through the area of maximum dynamic pressure of 720 pounds per square foot. Main engines had been throttled up to 104-percent thrust and the solid rocket boosters were increasing their thrust when the first flickering flame appeared on the right solid rocket booster in the area of the aft field joint. This first very small flame was detected on image-enhanced film at 58.788 seconds into the flight.

One film frame later from the same camera, the flame was visible without image enhancement. It grew into a continuous, well-defined plume at 59.262 seconds. At about the same time (60 seconds), telemetry showed a pressure differential between the chamber pressures in the right and left boosters. The right booster chamber pressure was lower, confirming the growing leak in the area of the field joint.

As the flame plume increased in size, it was deflected rearward by the aerodynamic slipstream and circumferentially by the protruding structure of the upper ring attaching the booster to the external tank. These deflections directed the flame plume onto the surface of the external tank. This sequence of flame spreading is confirmed by analysis of the recovered wreckage. The growing flame also impinged on the strut attaching the solid rocket booster to the external tank.

At about 62 seconds into the flight, the control system began to react to counter the forces caused by the plume and its effects. The left solid rocket booster thrust vector control moved to counter the yaw caused by reduced thrust from the leaking right solid rocket booster. During the next nine seconds, space shuttle control systems worked to correct anomalies in pitch and yaw rates.

The first visual indication that swirling flame from the right solid rocket booster breached the external tank was at 64.660 seconds when there was an abrupt change in the shape and color of the plume. This indicated that it was mixing with leaking hydrogen from the external tank. Telemetered changes in the hydrogen tank pressurization confirmed the leak. Within 45 milliseconds of the breach of the external tank, a bright sustained glow developed on the black-tiled underside of the Challenger between it and the external tank.

Beginning at about 72 seconds, a series of events occurred extremely rapidly that terminated the flight. Telemetered data indicate a wide variety of flight system actions that support the visual evidence of the photos as the shuttle struggled futilely against the forces that were destroying it.

At about 72.20 seconds the lower strut linking the solid rocket booster and the external tank was severed or pulled away from the weakened hydrogen tank permitting the right solid rocket booster to rotate around the upper attachment strut. This rotation is indicated by divergent yaw and pitch rates between the left and right solid rocket boosters.

At 73.124 seconds, a circumferential white vapor pattern was observed blooming from the side of the external tank bottom dome. This was the beginning of the structural failure of the hydrogen tank that culminated in the entire aft dome dropping away. This released massive amounts of liquid hydrogen from the tank and created a sudden forward thrust of about 2.8 million pounds, pushing the hydrogen tank upward into the intertank structure. At about the same time, the rotating right solid rocket booster impacted the intertank structure and the lower part of the liquid oxygen tank. The structures failed at 73.137 seconds as evidenced by the white vapors appearing in the intertank region.

Within milliseconds there was massive, almost explosive, burning of the hydrogen streaming from the failed tank bottom and the liquid oxygen breach in the area of the intertank.

At this point in its trajectory, while traveling at a Mach number of 1.92 at an altitude of 46,000 feet, the Challenger was totally enveloped in the explosive burn. The

Challenger's reaction control system ruptured and a hypergolic burn of its propellants occurred as [the vehicle] exited the oxygen-hydrogen flames. The reddish brown colors of the hypergolic fuel burn are visible on the edge of the main fireball. The orbiter, under severe aerodynamic loads, broke into several large sections which emerged from the fireball. Separate sections that can be identified on film include the main engine/tail section with the engines still burning, one wing of the orbiter, and the forward fuselage trailing a mass of umbilical lines pulled loose from the payload bay.

Evidence in the recovered wreckage from the 51-L mission supports this final sequence of events.

The Cause of the Accident. The concensus of the Commission and participating investigative agencies is that the loss of the space shuttle Challenger was caused by a failure in the joint between the two lower segments of the right solid rocket motor. The specific failure was the destruction of the seals that are intended to prevent hot gases from leaking through the joint during the propellant burn of the rocket motor. The evidence assembled by the Commission indicates that no other element of the space shuttle system contributed to this failure.

Last Words

On board all shuttle flights is a recorder for intercom communications. It records voice communications during countdown and launch that are not transmitted outside of the orbiter. This recorder was recovered from the ocean bottom after the accident and its recordings were restored.

The following is a transcript of what was said during those short 73 seconds. The words are mostly those of Commander Francis R. Scobee and Pilot Michael J. Smith. The numbers indicate elapsed seconds into the flight.

:00	Resnik: All right.
:01	Smith: Here we go.
:07	Scobee: Houston, Challenger roll program.
:11	Smith: Go, you mother.
:14	Resnik: LVLH [Cockpit switch change].
:15	Resnik: [Expletive] hot.
:16	Scobee: Ohhhhkaaay.
:19	Smith: Looks like we've got a lotta wind here today.
:20	Scobee: Yeah.
:22	Scobee: It's a little hard to see out my window here.
:28	Smith: There's 10,000 feet and Mach point five.
:30	[Garbled]
:35	Scobee: Point nine.
:40	Smith: There's Mach one.
:41	Scobee: Going through 19,000.
:43	Scobee: Okay, we're throttling down.
:57	Scobee: Throttling up.
:58	Smith: Throttle up.
:59	Scobee: Roger.
:60	Smith: Feel that mother go.
:60	Uncertain: Woooohoooo.
:62	Smith: 35,000 going through one point five.
:65	Scobee: Reading 486 on mine [airspeed].
:67	Smith: Yep, that's what I've got too.
:70	Scobee: Roger, go at throttle up.
:73	Smith: Uh-oh.

The Teacher in Space logo representing the space shuttle in flight carrying the first teacher into Earth orbit. The light of knowledge and education is symbolized by the flaming torch. Its message is to reach out, grasp the torch, share the learning experience, and then pass it on to future generations.

Teacher in Space finalists pose next to NASA's "zero-gravity" aircraft.

Sharon Christa McAuliffe, a high school social studies teacher in Concord, New Hampshire, accepts her selection as the first Teacher in Space as Vice President George Bush and Secretary of Education William Bennett observe. Announced by the Vice President in a White House ceremony, McAuliffe was chosen from over 10,000 applicants.

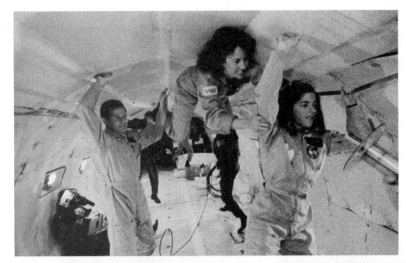

McAuliffe (center) shares a "zero-gravity" training flight with Congressman Bill Nelson (D-Florida) and Barbara R. Morgan. Morgan was McAuliffe's backup for 51-L and Nelson was due to fly on Mission 61-C.

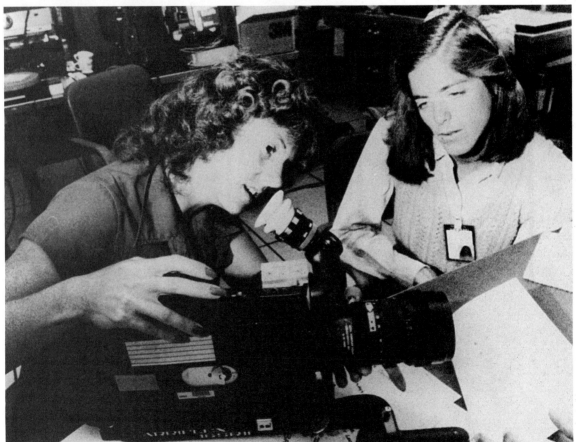

McAuliffe (left) and her backup, Barbara R. Morgan, began training September 30, 1985, by learning basic procedures for space travelers. The second week of training included camera training, aircraft familiarization, and other activities. Here, McAuliffe and Morgan get hands-on experience with an Arriflex motion picture camera following a briefing on space photography.

McAuliffe egresses the rear station of a NASA T-38 jet trainer at Ellington Air Force Base near the Johnson Space Center, where she was in training for the 51-L mission.

Here, five of 51-L's seven crewmembers receive slidewire escape training at Pad 39B during the Terminal Countdown Demonstration Test.

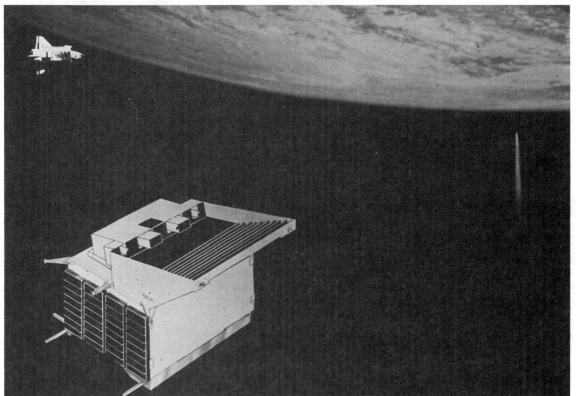

The University of Colorado's Spartan Halley spacecraft consisted of two ultraviolet spectrometers, attitude cameras, a microprocessor, and an elaborate baffling system, all mounted on a Spartan carrier developed by NASA's Goddard Space Flight Center. The service module held batteries, a tape recorder, an integrating gyrosystem, solar acquisition sensors, a star tracker, and jets of the attitude control system. Thermal louvers, on both sides of the Spartan carrier, controlled temperature. The Spartan Halley spacecraft was scheduled to be transported into space by 51-L and deployed into an independent orbit by the Remote Manipulator System. The spacecraft would have functioned autonomously as it observed Comet Halley. After approximately 48 hours, the shuttle would have rendezvoused with Spartan Halley, retrieved the spacecraft, and stored it once again in the payload bay.

In the foreground, shuttle Columbia is poised for future launch from Pad 39A. In the background Challenger sits atop the Crawler Transporter on its way to the newly reconfigured Pad 39B for the 51-L launch. This was to be the first time two orbiters were on launch pads at the same time.

January 28, 1986. An ice and frost inspection photograph taken prior to launch of 51-L.

Another ice and frost inspection photograph. The left solid rocket booster dominates the left half of the photo.

Challenger lifted off from Pad 39B on January 28, 1986 at 11:38 A.M. (EST). It was the first launch from Pad 39B in over 10 years.

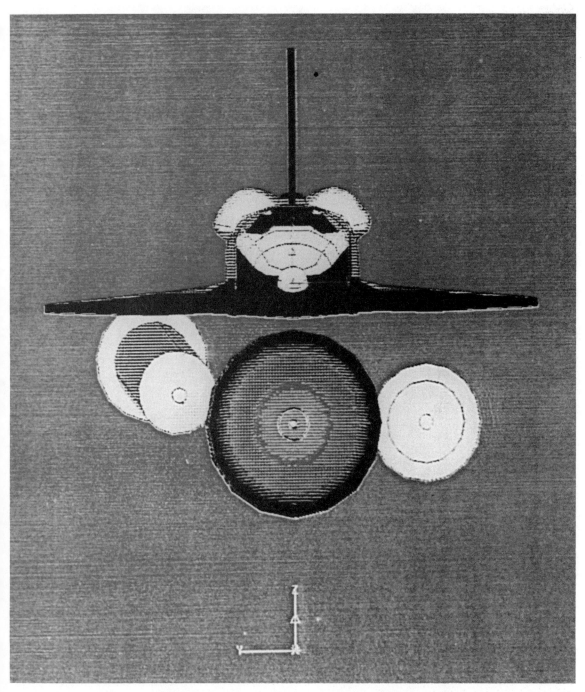

This computer-drawn photograph of a top view of the Challenger configuration suggests a solid rocket booster malfunction immediately prior to the explosion: the booster being released from its lower link to the external tank and left free to pivot about its forward attachment point and one of its remaining aft attachments, impacting the external tank's inner tank region between the liquid oxygen and liquid hydrogen tanks.

Wreckage from 51-L is retrieved from the Atlantic Ocean and returned to the tidal basin at Cape Canaveral Air Force Station aboard the USCG cutter Dallas on January 30.

A 9'7" by 16' segment of Challenger's right wing is shown after being off-loaded from the rescue and salvage ship USS Opportune. The piece of wreckage was located and recovered in mid-April 1986 by Navy divers from the Opportune about 12 nautical miles northeast of Cape Canaveral in 70 feet of water.

Family members of the Challenger 51-L crew meet with Vice President George Bush at the White House. Seated from left to right, Jane Smith, Cheryl McNair, Charles Resnik, Steven McAuliffe, and June Scobee.

Epilogue

Approximately one year after Challenger accident, NASA announced the flight crew for the 26th shuttle mission. The mission will be commanded by Frederick H. Hauck who led the 14th flight in November 1984 and was the pilot on the 7th flight in June 1983.

The pilot for the 26th flight will be Richard O. Covey. Covey served as pilot on the 20th flight in August 1985.

Mission specialists will be John M. Lounge, who flew as a mission specialist with Covey on the 20th flight; George D. Nelson, who served as a mission specialist on the 11th flight in April 1984 and the 24th flight in January 1986; and David C. Hilmers, who flew on the 21st flight in October 1985.

In announcing the crew, NASA Associate Administrator for Space Flight Richard H. Truly (himself a shuttle veteran) said, "The naming of the crew for the next flight is a major event in the process of returning the shuttle to flight. I am particularly pleased to assemble a group of such experienced individuals led by one of our senior spaceflight veterans, and I am very proud of them."

In late 1987, NASA announced the crewmembers for STS-27 (the 27th flight), a Department of Defense mission: Robert L. Gibson (Cdr., USN; pilot of STS-41-B and commander of STS-61-C), commander; Guy S. Gardner (Lt. Col., USAF), pilot; and Richard M. Mullane (Col., USAF; mission specialist on STS-41-D), Jerry L. Ross (Lt. Col., USAF; mission specialist on STS-61-B), and William M. Shepherd (Cdr., USN), mission specialists.

Index

Index

A
Abbey, George, **24**
abort mode switch, **209**
Abrahamson, Lt. Gen. James, 102, 120
ACCESS, 247, **248**, **254**, **256**
acoustic containment furnace, 108
Acton, Loren W., 202-215
Al-Saud, Sultan Salman, 188-201
Allen, Joseph P., 48-59, 152-159
alloy experiments, 101
Anik C-1 communications satellite, 166, 167, **170**
Anik C-2 communications satellite, 70-72, **78, 83**
Anik C-3 communications satellite, 48, 50, **51, 55**
Anik D-2 communications satellite, 152-154, 158, 159
animal enclosure module, 86-89, 108
animals aboard shuttle, 176-180, **184-186**
ant colony experiment, 72
apogee kick motor, **172**
Apollo 13, 16
Arabsat 1-B communications satellite, 188, 190, **196, 197**
Arnowitz, Leonard, 239
ASC-1 communications satellite, 216, 218, **219, 221**
Atlantis
  STS-51J, 230-235
  STS-61B, 244-257
atmospheric gas experiments, 101
atmospheric physics experiments, 202, 205
aurora australis, **183**
AUSSAT-1 communications satellite, 216, 218, **219**, 246

AUSSAT-2 communications satellite, 244, 246, **248**
automatic space experiment, 72
auxiliary power units, 27

B
Bartoe, John-David, 202-215
Baudry, Patrick, 188-201
Beggs, James M., **22,** 29, 41, 63
Bennett, Sect. William, 271
Bergman, Jules, **20**
biofeedback experiments, 86, 87
biomedical experiments, 188
Bluford, Guion S., 86-97, 236-243
Bobko, Karol J., 60-69, 166-175, 230-235
Bolden, Charles F. Jr., 258-265
booster recovery ships, 34
Booth, John R., 162
Brand, Vance, 48-59, 106
Brandenstein, Daniel C., 86-97, 188-201
Bridges, Roy D., 202-215
Buchli, James F., 160-165, 236-243
Bush, George, **6, 13**, 271, 283

C
capillary pump loop experiment, 192
capsule communicators (CAPCOMs), **22**
Cenker, Robert, 258-265
Cepollina, Frank J., 120
Challenger
  improvements to, 61-62
  payload bay of, **67, 91**
  satellite deployment from, **80**
  STS-6, 60-69
  STS-7, 70-85
  STS-8, 86-97
  STS-41B, 106-115

STS-41C, 116-127
STS-41G, 142-151
STS-51B, 176-187
STS-51F, 202-215
STS-51L, 266-284
STS-61A, 236-243
Challenger tragedy, v, 266-284
  cause of accident in, 269
  ice and frost buildup in, **276, 277**
  Rogers Commission report on, ix, 267
  unusual plume from solid rocket booster in, **279**
  wreckage retrieval from, 282
Chang-Diaz, Franklin, 258-265
Cleave, Mary L., **75**, 244-257
closed loop test, 204
Coats, Michael L., 128-141
Columbia
  auxiliary power unit overheating for STS-3, 27
  cargo bay of, **11, 55, 125**
  fuel cell failure during STS-2, 16
  malfunctions before STS-2, 15-16
  overhaul of, 259
  STS-1, 1-13
  STS-2, 14-25
  STS-3, 26-37
  STS-4, 38-47
  STS-5, 48-59
  STS-9, 98-105
  STS-61C landing trouble with, 261
  STS-61C, 258-265
conductivity experiments, 101
contamination monitor package, 88
contingency water landing, **5**
Continuous Flow Electrophoresis System, 38, 40, 72-73, 86-89, **94,** 128-130, 244, 247
convection experiment, **54**

289

coronal helium abundance spacelab experiment (CHASE), 205
cosmic ray upset experiment, 88
Covey, Richard O., 34, 216-229, 285
crawler transporter, **44**
Creighton, John, 188-201
Crippen, Robert L., 1-13, 70, 116-127, 142-151
crystals
   formation experiments with, 51, **53**, 101, 192
   organic, formation of, 247, 251

D

Department of Defense
   first fully dedicated flight to, 160, 161, 162
   payloads form, 38
developmental flight instrumentation package, **11**, 86, 87, **91**
diffusive mixing of organic solutions, 247
direct insertion launch technique, 116, 117
Discovery
   STS-41D, 128-141
   STS-51A, 152-159
   STS-51C, 160-165
   STS-51D, 166-175
   STS-51G, 188-201
   STS-51I, 216-229
Draughon, Harold, 41
drop dynamics module, **187**
dual cycle high-specific-impulse liquid propulsion system, vii
Dunbar, Bonnie J., **75**, 236-243

E

earth radiation budget experiment (ERBE), 145, 146
Earth survey, STS-2, 17
EASE, 244, 246, 247, **248, 255**
echocardiograph experiments, 190
Edwards Air Force Base, first hard landing at (STS-4), 41
ejectable plasma diagnostics package (PDP), 206
electrophoresis test, 26, 28, **31, 37**, 40, 108
   commercial, 38
elemental composition and energy spectra of cosmic rays experiment, 206
emergency egress training, **65**
England, Anthony W., 202-215
Engle, Joseph H., 14-25, 216-229
Enterprise prototype, **233**
European Space Agency, 108
   first representative aboard shuttle from, 98

joint experiments by NASA and, 98, 99
Spacelab 2 experiments for, 205
STS-51B laboratory flight for, 176, 177, 178
STS-9 photographic experiments for, 102
extravehicular activity power tools, 116, 117

F

Fabian, John M., 70, 188-201
far ultraviolet space telescope, 102
feature identification and location experiment (FILE), 145
female astronauts, **75**
first American black astronaut (STS-8), 86, 87
first American woman in space, 70
first commercial mission specialist, 128, 129
first night launch and landing (STS-8), 86-87
first non-American astronaut in shuttle, 98
first non-astronaut scientist in space, 98
first student space experiment, 26, 28, 31
Fisher, Anna L., **75**, 152-159
Fisher, William F., 216-229
FleetSatCom satellites, 219
flight objectives
   STS-1: safe launch, performance, and return, 1
   STS-2: remote manipulator system testing, 14
   STS-3: first student space experiment, 26, 28
      remote manipulator system test, 26
      space science experiments, 26, 28
      thermal response of orbiter, 26
   STS-4: continuous flow electrophoresis system, 38
      Defense Department payload, 38
      final research and development flight, 38
      induced environmental contamination monitor, 38
      zero-g medical research, 38
   STS-5: communication satellite deployments, 48
      first use of mission specialists, 48
   STS-6: first shuttle space-walk, 60
      tracking and data relay satellite deployment, 60
   STS-7: first American woman in space, 70
      largest crew, 70
      satellite deployment and retrieval, 70
      space adaptation syndrome experiments, 70
   STS-8: first American Black astronaut, 86
      first night launch and landing, 86
      large payload deployment testing, 86
      satellite deployment, 86
   STS-9: first non-American aboard, 98
      first non-astronaut scientist aboard, 98
      joint experiments of NASA and Eur. Sp. Agency, 98
      Spacelab 1 experiments, 98
   STS-41B: manipulator foot restraint used, 106
      manned maneuvering unit used, 106
      satellite deployments, 106
      shuttle pallet satellite reflown, 106
      untethered space-walks, 106
   STS-41C: direct insertion launch technique, 116
      extravehicular activity power tools used, 116
      in-orbit capture, repair, redeployments, 116
      long duration exposure facility deployed, 116
      manipulator foot restraint used, 116
      operational use of MMU, 116
   STS-41D: first commercial mission specialist aboard, 128
      heaviest payload, 128

STS-41G: largest spacecraft structure deployed, 128
multiple satellite deployments, 128
first satellite refueling, 142
first woman spacewalker, 142
large format camera use, 142
largest flight crew, 142
MAPS experiment, 142
Moceanic phenomena experiments, 142
shuttle imaging radar used, 142
STS-51A: multiple satellite deployment and retrieval, 152
STS-51B: animals as part of flight crew, 176
European Space Agency laboratory flight, 176
microgravity materials-processing and fluid experiments, 176
STS-51C: first fully Department of Defense flight, 160
inertial upper stage (IUS) booster deployed, 160
STS-51D: satellite deployments, 166
STS-51F: Spacelab 2 deployment, 202
STS-51G: international satellite deployments, 188
STS-51I: satellite deployment, retrieval, repair, 216
spacewalk, 216
STS-51J: Department of Defense flight, 230
STS-61A: German Spacelab D-1 mission, 236
STS-61B: erecting structures in space, 244
multiple satellite deployments, 244
STS-61C: Halley's Comet research, 258
satellite deployment, 258

flight readiness firing, **232**
flight support structure, **120**
flight verification test canister, 30

fluid motion experiments, 101
fly-by-wire digital computers, vii
Friendship 7, v
fuel tank, **42**
Fullerton, Charles G., 26-37, 202-215
Furrer, Reinhard, 236-243

G

Gardner, Dale A., 86-97, 152-159
Gardner, Guy S., 285
Garn, Sen. E.J. "Jake", v, 166-175
Garneau, Marc, 142-151
Garriott, Owen, 98
Gemini 5, 16
German Spacelab D-1, 236, 237, 238
experiments conducted aboards, 238
payload elements of, 238
Germany, Dan, 132
Getaway Specials, 28, 38, 40, **43**, 86, 87, 108, 188, 191, 238, 247
flight verification test canister, **30**
mixing molten metals, 51
Gibson, Robert L., 106, 258-265, 285
Glenn, John, v
global low orbiting message relay satellite (GLOMR), 179, 180, 238, 239
Grabe, Ronald J., 230-235
gravity influenced lignification experiment, 207, **215**
Gregory, Frederick D., 176-187
Griggs, S. David, 166-175

H

Halley's Comet research, 258, 259, 261, 266
hard x-ray imaging, 206
Hart, Terry J., 116-127
Hartsfield, Henry W., 38-47, 128-141, 236-243
Hasselblad camera, 199
Hauck, Frederick H., 70, 152-159
Hauersperger, Karla, 40
Hawley, Steven A., 128-141, 258-265
heads up display, 62
heflex bioengineering test, 26
helium-cooled infrared telescope, 207
Henize, Karl G., 202-215
high precision tracking experiment, 192
high-energy astrophysics experiments, 202, 206
Hilmers, David, 230-235, 285
Hoffman, Jeffrey A., 166-175

honeybee experiment, 119, **126**
Hurrican Elena, **226**

I

IMAX motion picture camera, 128, 129, **139**
induced environment contamination monitor, 26, 38, 41, **45**
inertial upper stage (IUS) booster rocket, 160-162
infrared astronomy experiments, 202, 206
INSAT-1B communications satellite, 86-89, **90, 91, 95**
Instrument Pointing system, 204, 205
ionospheric experiments, 101
Issel, Michelle, **53**

J

Jarvis, Gregory, 266-284

K

Ku-band antenna, 72, 86, 87, 88
Kusske, Amy, 40

L

large format camera, 142, 145, **155**
largest flight crew, 142
Leasat 2 communications satellite, 130, **135**
Leasat 3 communications satellite, 166
retrieval and repair of, **222, 223**
STS-51D in-orbit repairs to, 167, 171
STS-51I retrieval and repair of, 216, 217, 218
Leasat 4 communications satellite, 218, **219, 222**
Leestma, David C., 142-151, 207
Lenoir, Dr. William B, 48-59
Lewis, Dr. Marian, **31**
Lichtenberg, Dr. Byron K., 98
life sciences experiments, 100, 202, 207, 236
Lind, Don L., 176-187
liquid mercury experiment, 72
liquid sloshing behavior experiment, 191
long duration exposure facility, 116, 117, 119, **120, 121**
Lounge, John M., 216-229, 285
Lousma, Jack R., 26-37
Lovelace, Dr. Alan, **13**
Lucid, Shannon, **75,** 188-201

M

manganese-bismuth studies experiment, 191
manipulator foot restraint, 106, 107, 116-118, **124**, 159

291

manned maneuvering unit, 106-109, **110, 113, 114,** 118, **122, 172**
   operational use of, 116, 117
Marangoni convection, 101
materials-processing experiments, 100, 101, 188, 236
Mattingly, Thomas K., 38-47, 160-165
McAuliffe, Christa, 266-284
McBride, Jon A., 142-151
McCandless, Bruce, 106
McNair, Ronald E., 106, 266-284
measurement of air pollution from satellites (MAPS) experiments, 142, 145
Merbold, Dr. Ulf, 98
Messerschmid, Ernst, 236-243
metal-mixing experiment, 51
microgravity research experiments, 176, 177, 236
microwave scanning beam landing system, 17
mirror fabrication experiment, 247
mission specialists, STS-5 as first use of, 48
monodisperse latex reactor, 26, 28, 40, 72, 73, 108
Moore, Gilbert, 40
Moore, Jesse, 146, 193, 203
Morelos 1 communications satellite, 188, 190, **196**
Morelos-B communications satellite, 244, 246, **248, 253**
Morgan, Barbara R., 272
Mullane, Richard M., 128-141, 285
Musgrave, Dr. F. Story, 60-69, 202-215

N

Nagel, Steven, 188-201, 236-243
National Science Teachers Association, 28, 40, 51
   Shuttle Student Involvement Project, 28
National Space Shuttle Student Involvement Project, 31
Nelson, Dr. George D., 116-127, 258-265, 285
Nelson, Rep. Bill, 1, 258-265
Nelson, Todd E., 28, **31**
Neri, Rodolfo, 244-257
Northern Utah Satellite (NUSAT), 178

O

oceanic phenomena experiments, 142
Ockels, Wubbo, 236-243
Onizuka, Ellison S., 160-165, 266-284
orbital maneuvering system (OMS) pods, 11

orbital maneuvering system engine, **83**
Orbiter Processing Facility, **103**
OSTA-2 payload, 71, 73
Overmyer, Robert F., 48-59, 176-187
oxygen glow phenomenon, 89, 90, **95**
O'Connor, Bryan D., 244-257

P

Pailes, William A., 230-235
Palapa B-1 communications satellite, 71, 72, **78, 83**
Palapa B-2 communications satellite, 106-108, **110, 111**
   STS-51A retrieval and repair of, 152-154, 158, 159
Parker, Dr. Robert A., 98
particle analysis cameras for the shuttle (PACS), 261
payload assist module, 50, 72, **90,** 108, 245
payload flight test article (PFTA), 86-89, **91, 96**
payload operation control center, 204
Payton, Gary E., 160-165
Peterson, Donald H., 60-69
pharmaceutical manufacture experiments, 130, 131
Pioneer, v
plant growth units, **215**
plasma depletion experiments, 206
plasma diagnostics package (PDP), 29, **211, 212**
plasma physics experiments, 202, 206
postural experiments, 190

R

Reagan, Pres. Ronald, **13, 22,** 63, 90
recoverable spaceflight vehicle, development of, vii
remote manipulator system, 16, 22, 29, 38, **45, 80-83, 91, 124, 158, 170**
   icing on, 131, 132
   large payload test for, 86-89
   payload released by, 73
   testing of, 14, 26
research animal holding facility, **184**
Resnik, Judith A., **75,** 128-141, 266-284
Richards, Dick, 132, 146, 193
Ride, Dr. Sally K., v, **22,** 70, **75,** 142-151
Rogers Commission, 267
Rogers, Sect. William P., ix, 267
Ross, Jerry L., 244-257, 285

S

Satcom Ku-1 communications satellite, 258, 259, **264**
Satcom Ku-2 communications satellite, 245, **253**
satellite
   deployments of, 48, 50, **57**
   failed deployment of, 108
   in-orbit capture, repair, redeployment of, 116, 117
   multiple deployment and retrieval/repair of, 152-154, 158, 159
   retrievals and proximity operations, 70, 72
   STS-41G refueling experiments with, 142, 146
   STS-41D multiple deployments of, 128, 129, 130
   STS-51D deployment of, 166
SBS-3 communications satellite, 48, 49, **51, 55, 57**
SBS-4 communications satellite, 130, **135**
Scobee, Francis R., 116-127, 266-284
Scully-Power, Paul, 142-151
Seddon, M. Rhea, **75,** 166-175
seed germination experiments, 72, 192
Shaw, Brewster H. Jr., 98, 244-257
Shepherd, William M., 285
Shriver, Loren J., 160-165
shuttle carrier aircraft, **25**
shuttle entry air data system, 260
shuttle imaging radar-B (SIR-B), 143-145
shuttle infrared leeside temperature sensing experiment, 260
shuttle pallet satellite (SPAS-01), 70-73, **80-83, 110**
   STS-41B reuse of, 106, 107
Shuttle Student Involvement Project, 40, 51
Sieck, Robert, 162
Slayton, Donald K. "Deke", **20**
slipcasting experiment, 191
small self-contained payload program, 40
Smith, Michael J., 266-284
snow-crystal growth experiment, 88
snowflower germination experiment, 72
solar array panel deployment, 128-131, **136, 137**
solar magnetic and velocity field measurement system, 205
Solar Maximum Mission satellite, viii, 109, **123**
   STS-41C repair of, 117-119
solar optical universal polarimeter (SOUP), 205, **210**

solar physics experiments, 202, 205
solar ultraviolet high resolution telescope and spectrograph (HRTS), 205
solar ultraviolet spectral irradiance monitor (SUSIM), 206
soldering operations experiment, 72
solid rocket booster
    Challenger explosion of, 279
    dewatering of, **8**
    recovery of, **9**
space adaptation syndrome, 70, 73, 86, 87, 90, 98, 100, 168
space experiments with particle accelerators (SEPAC), 101
space shuttle
    cutaway view of, **4**
    remote manipulator system for, **80**
    thermal protection system for, **35**
space sickness, 73, 86, 87, 90, 100, 168
Space Transportation System, ix
space ultraviolet radiation experiment, 192
Spacelab 1, 98-101, **103**
Spacelab 2, 202-204, **210**
    scientific experiments aboard, 205
Spacelab 3, 176-179
Spacelab pallet, 17
spacewalk
    STS-6, 60, 63, **69**
    STS-41G, first woman in, 142, 146
    STS-51I, 216
    STS-61B, 244
    untethered, STS-41B, 106, 107, 109
Spartan 1 communications satellite, 188, 192, **198**
Spartan Halley experiment, 266, **274**
sponge growth experiment, 51
Spring, Sherwood C., 244-257
Stewart, Robert, 106, 230-235
Stone, Randy, 132
stowage assembly box, 110
strategic defense initiative experiments, 188, 192
STS-1: Columbia, 1-13
STS-2: Columbia, 14-25
STS-3: Columbia, 26-35
STS-4: Columbia, 38-47
STS-5: Columbia, 48-59
STS-6: Challenger, 60-69
STS-7: Challenger, 70-85
STS-8: Challenger, 86-97
STS-9: Columbia, 98-105
STS-26, 285
STS-27, 285

STS-41B: Challenger, 106
STS-41C: Challenger, 116-127
STS-41D: Discovery, 128-141
STS-41G: Challenger, 142-151
STS-51A: Discovery, 152-159
STS-51B: Challenger, 176-187
STS-51C: Discovery, 160-165
STS-51D: Discovery, 166-175
STS-51F: Challenger, 202-215
STS-51G: Discovery, 188-201
STS-51I: Discovery, 216-229
STS-51J: Atlantis, 230-235
STS-51L: Challenger, 266-284
STS-61A: Challenger, 236-243
STS-61B: Atlantis, 244-257
STS-61C: Columbia, 258-265
student space experiments, 51, 71, 191, 192
Sullivan, Kathryn D., **75**, 142-151, 142
superfluid helium experiments, 207
surface tension experiments, 51, 101
Syncom IV-4 communications satellite, **135**, 218, **219**

T
T-38 chase plane and trainer, **34, 76, 273**
Teacher-in-Space project, 266
technology research experiments, 202, 207
Telesat H communications satellite, 152-154, 158, 159
Telstar 3 communications satellite, 130, 188, 190, 191, **197**
Thagard, Dr. Norman E., 70, 176-187
thermal protection (tile) system, vii, **35, 42**
thermal response of orbiter, 26, 29, 39, 89
Thomas, Scott, **54**
Thornton, Dr. William E., 86-97, 176-187
tilt table, **67**
tracking and data relay satellite, 86-89, 145
    ground communications through, 72
    STS-6 deployment of, 60, 62, **68**
    STS-51L uncompleted deployment of, 266
Truly, Richard H., 14-25, 86-97, 285
Turner, Gary, 131
Typhoon Pat, **227**

U
ultraviolet photographic experiment, 88
untethered spacewalk, STS-41B, 106, 107, 109
upper atmosphere mass spectrometer, 260
Urban, Dr. Eugene, 205, 207
UTC Liberty booster recovery ship, **34**

V
van den Berg, Lodewijk, 176-187
van Hoften, Dr. James D., 116-127, 216-229
vehicle charging and potential experiment (VCAP), 206
very wide field camera, 102
vitamin D metabolites and bone demineralization experiment, 207
Voyager, v

W
Walker, Charles D., 128-141, 166-175, 244-257
Walker, David M., 152-159
Wang, Taylor G., 176-187
weightless environment training facility, **52**
Weinberger, Caspar, 162
Weitz, Paul J., 60-69
Westar IV communications satellite, **110-111**
    STS-51A retrieval and repair of, 152-154, 158, 159
Westar VI communications satellite, 106, 107, **172**
    failed deployment of, 108
Williams, Donald E., 166-175

X
x-ray imaging, 206
x-ray mapping experiments, 192
x-ray spectroscopy experiments, 102

Y
Young, John W., 1-13, 63, 77, 98

Z
zero-g medical research, 40

Edited by Carl H. Silverman

Other Bestsellers From TAB

☐ **U.S. CIVIL AIRCRAFT SERIES, VOL 9: ATC 801—ATC 817**—Joseph P. Juptner

This intriguing volume covers the last 17 Approved Type Certificated Aircraft plus a master index of the nine volume series, ATC update, group two section (Letter of Approval), limited type certificate (LTC), restricted category (AR), people of aviation index. Included are plane descriptions, histories, production and performance data, complete specifications, and other technical information. 240 pp., illustrated.

Hard $19.95   Book No. 29182

☐ **THE HUNGRY TIGERS: THE FIGHTER PILOT'S ROLE IN MODERN WARFARE**—Frank J. O'Brien

Now, a former fighter pilot whose impressive service record includes 163 combat missions in Vietnam, provides a fascinating insight into the realities of today's U.S. Tactical Air Forces. O'Brien puts his emphasis on the human element—the "hungry tigers" whose pride, dedication, and sharply-honed skills set them apart from ordinary service men. Includes vivid descriptions of the modern fighter plane and those of the future. 320 pp., 78 illus.

Paper $14.95   Book No. 22395

☐ **ACES OVER THE OCEANS—THE GREAT PILOTS OF WORLD WAR II**—Edward H. Sims

This exceptional account gives you detailed, dramatic, and colorful coverage of the 12 most dramatic naval air battles that took place over the seas during WWII. Ed Sims, a former pilot who served in the Eighth Air Force in Europe, has compiled these stories as a legacy to the heroism of the naval pilots of the United States, Germany, and Great Britain. Personal interviews that Sims conducted with the actual pilots who flew these missions set the stage for a dramatic account of these battles that will keep you on the edge of your seat. 200 pp., 22 illus., 8 page B&W photo section.

Paper $14.95   Book No. 22392

☐ **STUDIES IN STARLIGHT: UNDERSTANDING OUR UNIVERSE**—Charles J. Caes

Man, for all his intelligence and technology, has yet to understand the power of radiant energies . . . or perhaps even to discover all of them. Even those with only limited exposure to electromagnetic concepts will find this book engrossing—and understandable. Pictures and prose relate the histories of the efforts made to understand the mysteries of our universe. This is a book that belongs in the collections of scientists and star gazers alike. 256 pp., 133 illus.

Paper $12.95   Hard $18.95
Book No. 2946

☐ **THE HELICOPTER**—Keith Carey

Veteran helicopter pilot Keith Carey has produced a fascinating book that provides a detailed description of the evolution of the helicopter from the first "semi-successful" versions to today's sophisticated machines. Meticulously researched and illustrated with more than 200 photographs and diagrams of how helicopters work, this is a reference that should not be missed by anyone fascinated with rotorcraft in particular, or aviation history in general. 224 pp., 227 illus.

Paper $14.95   Book No. 2410

☐ **WINGS OF THE WEIRD AND WONDERFUL**—Captain Eric Brown

A fascinating assessment of the flying qualities of unusual and outstanding aircraft. This is the author's seventh book on aviation and has been acclaimed as a classic among aviation buffs worldwide. Filled with pictures, you'll see the Kingcobra in flight, the Grumman Cougar on its land catapult trials, the De Havilland Sea Mosquito landing on an aircraft carrier—the first twin-engine ever to do so, and more! 176 pp., 77 illus.

Paper $12.95   Hard $19.95
Book No. 2404

*Prices subject to change without notice.

Look for these and other TAB books at your local bookstore.

TAB BOOKS Inc.
Blue Ridge Summit, PA 17294-0850

Send for FREE TAB Catalog describing over 1200 current titles in print.
OR CALL TOLL-FREE TODAY: **1-800-233-1128**
IN PENNSYLVANIA AND ALASKA, CALL: **717-794-2191**